All images in this book are from the archives of NASA captured by HiRISE, licensed under CC0 1.0.

Images are all color-enhanced and presented unrelated with clear and engaging facts that are as enlightening as they are breathtaking.

Text, book, and cover design © 2023 by Jehan Radwan

PLANET MARS

UP CLOSE LIKE NEVER BEFORE

TEXT BY JEHAN RADWAN

Mars has been a fascination to humanity since the discovery of its existence. Being our neighboring planet, glancing in the sky with its signature Red hue and so similar in size to the Earth, we lately started wondering if it possibly could be our second home.

In 1976, NASA's orbiter Viking 1 took the first picture of Mars in black and white, representing a milestone at the time. Since then, we have come a long way. Thanks to NASA's Sojourner (1997), Spirit (2004–2010), Opportunity (2004–2018), Curiosity (2012–present), and Perseverance (2021–present) rover, along with China's Zhurong (2021-2022), we are now able to discover all its beauty up close!

ENIGMATIC REALMS

HYPNOTIC ATTRACTION

Dive into the mysteries of Mars and see the planet with its impressive landscapes up close as it reveals its beauty beyond our imagination. Discover astonishing facts about our distant planet that is able to produce its own glowing light apart from the sun!

Although the following images have been color-enhanced, their celestial landscapes are breathtaking. Mars surprises us with mesmerizing formations and panoramas that are mysterious and spectacular in a raw and artistic way.

Page 4 | Mesmerizing depositions and eroding bedrocks captured in a central pit located in the Schaeberle Crater.
Page 5 | Charming dunes captured in Nili Patera on top of a large lava bed from an ancient volcano.
Page 7 | Image taken in Syrtis Major – often referred to as the "dark spot." The Planum's dark color comes from the basaltic volcanic rocks in this area and was the first documented surface feature of another planet in 1659.

Page 9 | Beautiful formations of dunes located in Vastitas Borealis, the north polar region on Mars. It is not officially recognized as a feature.
Page 13 | Captured clear layering in the sand in the southern Holden crater. Such scenes represent a window into Mars's complicated geological past and history.

Ever since the beginning of its discovery, Mars has spiked the imagination of humanity. Sceneries with a booming fantasy got picked up in literature and later in movies, such as The War of the Worlds, viewed by millions of people, manifesting fantasies about green Martian aliens over the last centuries.

Our solar system holds eight planets and one star, the Sun. If we look at the other planets, it quickly becomes clear why Mars is the focus of our interest.

Our outer planets, Neptune, Uranus, Saturn, and Jupiter, are all Jovian Planets, gas giants without any solid surface. The remaining planets are called Terrestrial as they have compact and rocky surfaces like Earth's "Terra Firma."
Looking at Mercury, the closest planet to the Sun, the planet holds an average temperature of 167°C (333°F), facing a challenging environment for any man-made material, let alone even life, to be considered a planet to visit.

However, the planet got orbited by NASA's Mariner 10, covering about 45% of the surface in images. Additionally, NASA's Messenger flew by the planet three times and orbited Mercury for four years.

Attempts were made to send a lander in a study in 2010, but no mission ever took place as the planet presented itself as such a challenging target.

While Venus is the closest planet in orbit, it has been evident since the late 1950s that the planet would be too hot to be ever considered a place to visit by astronauts, with an average of 751 Kelvin converting to 477°C (892°F) due to a dense CO2 atmosphere.
The longest any spacecraft has survived is a little over two hours – set by the Soviet Union's Venera 13 in 1981.

The remaining planet, Mars, is the closest in orbit to Earth away from the Sun, with an average temperature of -65°C (-85°F). Its conditions raise questions of whether there is life, was there life, or can we live there?

Page 15 | Gorgeous Landscapes captured above Arabia Terra. The land holds many interesting features to discover more about Mars's geology.
Page 17 | Hummock in the Iani Region, often referred to as "Iani Chaos." It is speculated to have formed below the surface of water or ice.
Page 20 | Caption from the southern polar region of Mars. The cap is permanent and consists of dry ice and solid carbon dioxide. The image shows mesmerizing pits filled with condenses on flat areas.
Page 21 | Split soil by alluring Polygons. These artistic shapes are formed due to dry ice – shaping Mars's surface and contributing to its breathtaking beauty.

ANCIENT CHRONICLES

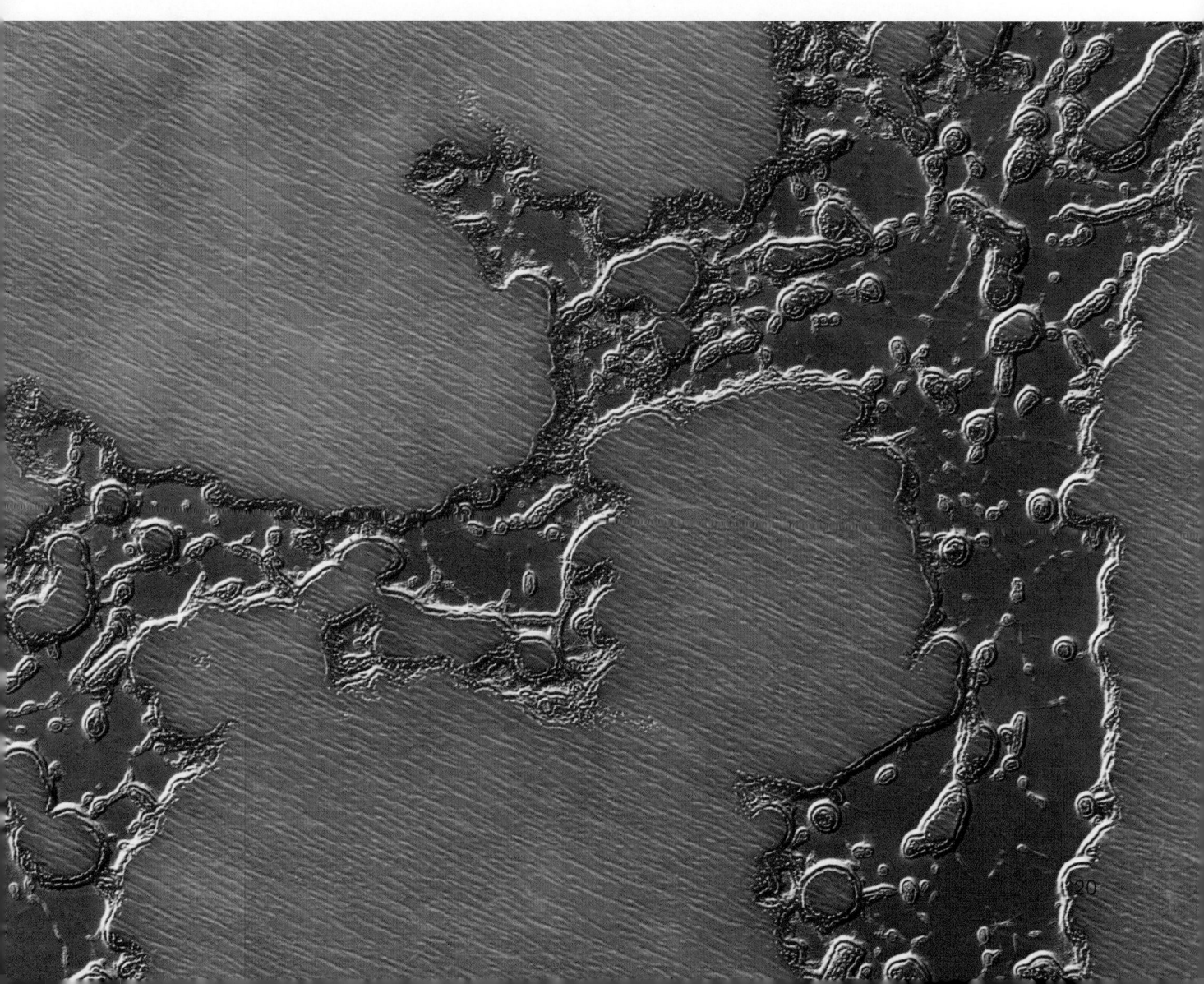

Detailed knowledge from all our neighboring celestial objects is compared to the discovery of its existence very young. All planets visible to the naked eye up to Saturn are well documented through famous astronomers.

The first written record of Mars dates back to ancient Egyptian astronomers around the 2nd millennium BCE. "Her Desher," Mars's ancient name given by the Egyptians, translates to "The red one." They were aware of his retrograding motion as the planet appears to move backward every two years, 26th month to be exact. An illusion due to the relative positions of the planet and Earth and how they are moving around the Sun.

MOLECULAR WATER

Galileo Galilei was the first astronomer in 1610 who was able to take the first accurate observation of the planet through his own telescope.

Despite the belief that Galilei had invented the telescope, he did redesign and build his own with higher magnifying power for his own use.
The actual first telescope was invented in the Netherlands with a patent record dated the 2nd October 1608 filed by Hans Lippershey for his instrument "seeing things far away as if they were nearby."

———————————

Page 24 | These exquisite dunes and ripples have been captured in the Southern hemisphere of Mars.
Page 25 | Hellas Planitia is a field located deeper than most of Mars's dune fields at this latitude. Latitude refers to positions horizontally on a planet.
Page 27 | A densely cracked ground between beautiful layers captured in the north of Arabia Terra.

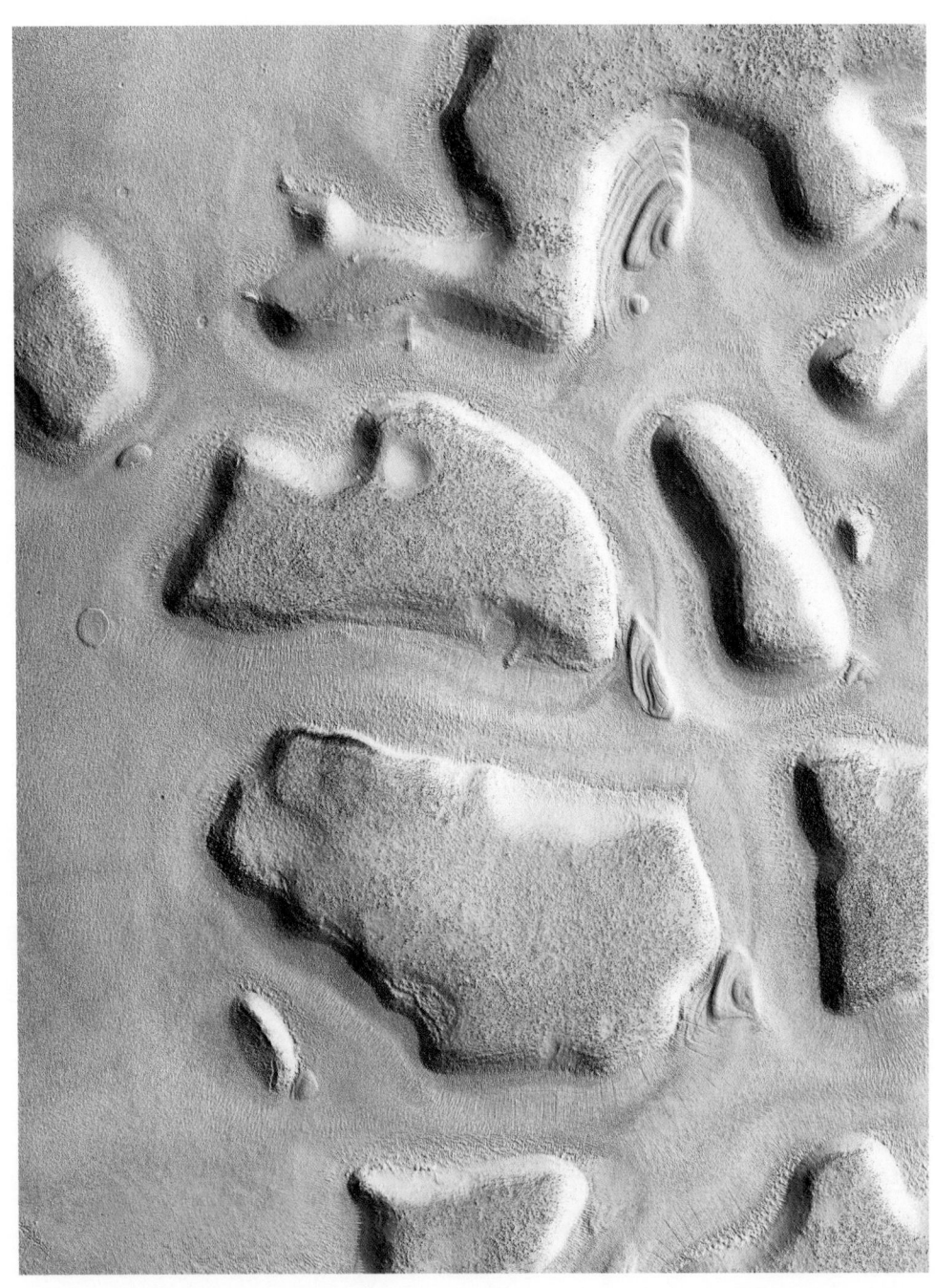

Page 30 – Tectonic shifts or volcanic activity can create pits like these. They are of great geological interest and provide direct windows into the planet's history. This beautiful example was captured on Arsia Mons.
Page 31 - Gorgeous layered formations located in Utopia Planitia. Utopia is the largest and most significant impact basin on Mars.
Page 33 – Sand rivers and cracks between dust plains captured in the Elysium Planitia region.

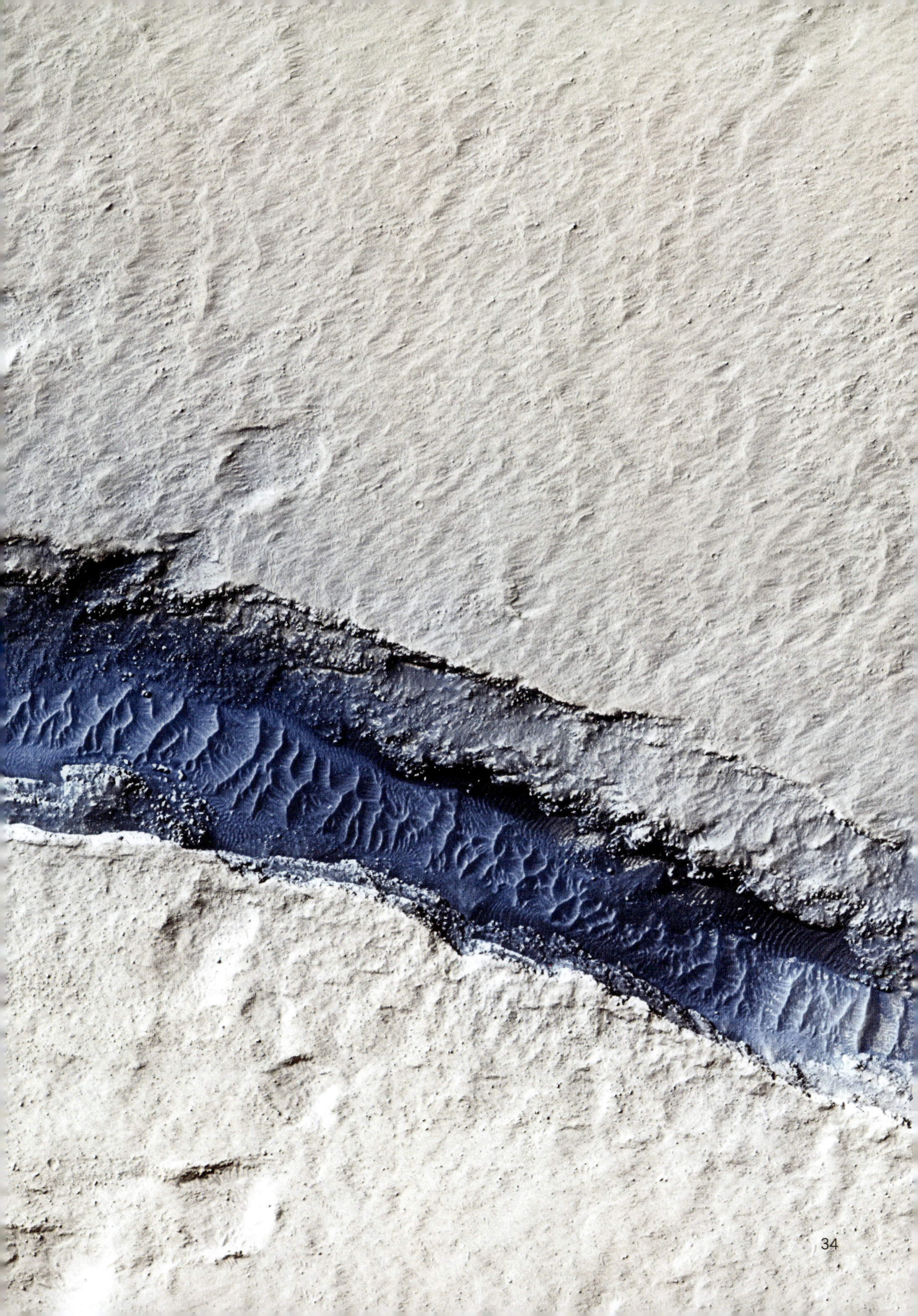

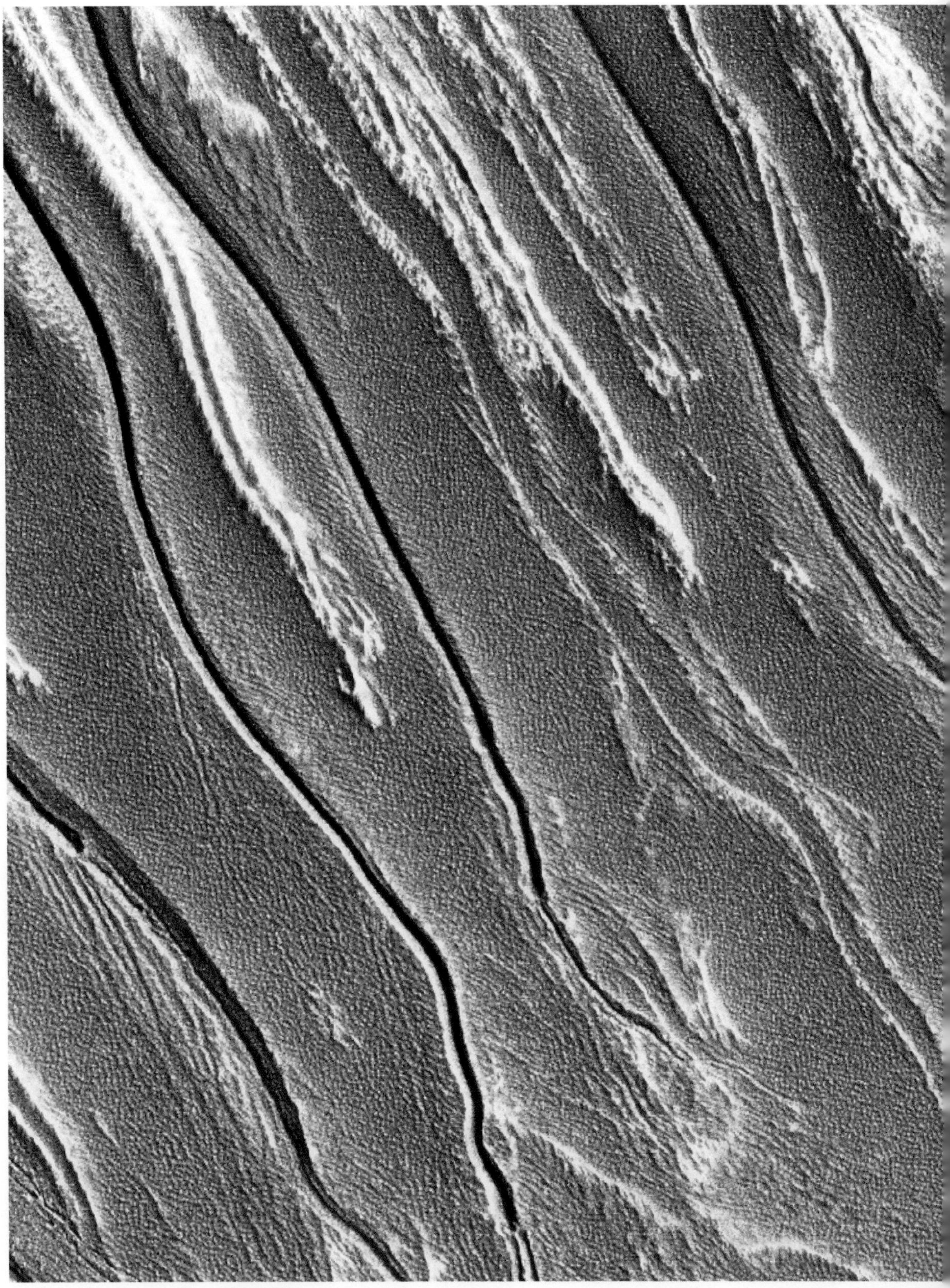

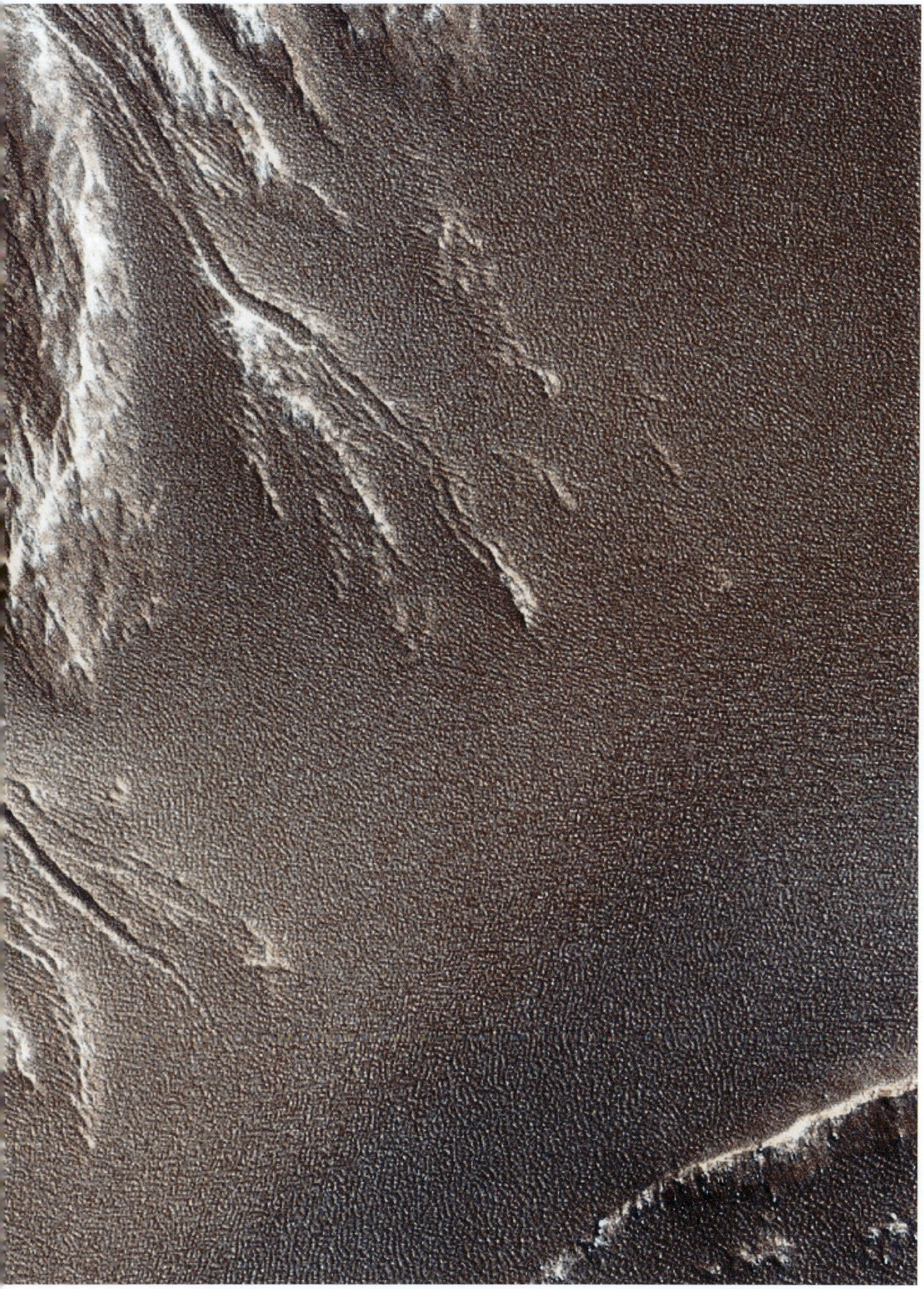

Page 35 - Deep gullies on crater walls. Their formation is still up for debate, but it is widely believed they have been formed due to carbon dioxide frost.
Page 39 – Mesmerizing frost in dune gullies in the Matara Crater. Such formation usually occurs when seasonal frost is present.

Superior observations made in 1999 by NASA on the red planet Mars is the discovery of traces of water, a vital component of our planet, is not a guarantee on other celestial planets.

The detection of water vapor on Mars, came as a significant revelation, adding to our understanding of the planet's past and its potential.

An ever-present mantra has been "follow the water" in humanity's persistent search for life in the universe. We now have convincing science that validates its presence.

Was there ever life?

Scientists found solid evidence of liquid water in 2015, confirming the evidence of liquid water flowing beneath the ice cap on the red planet.

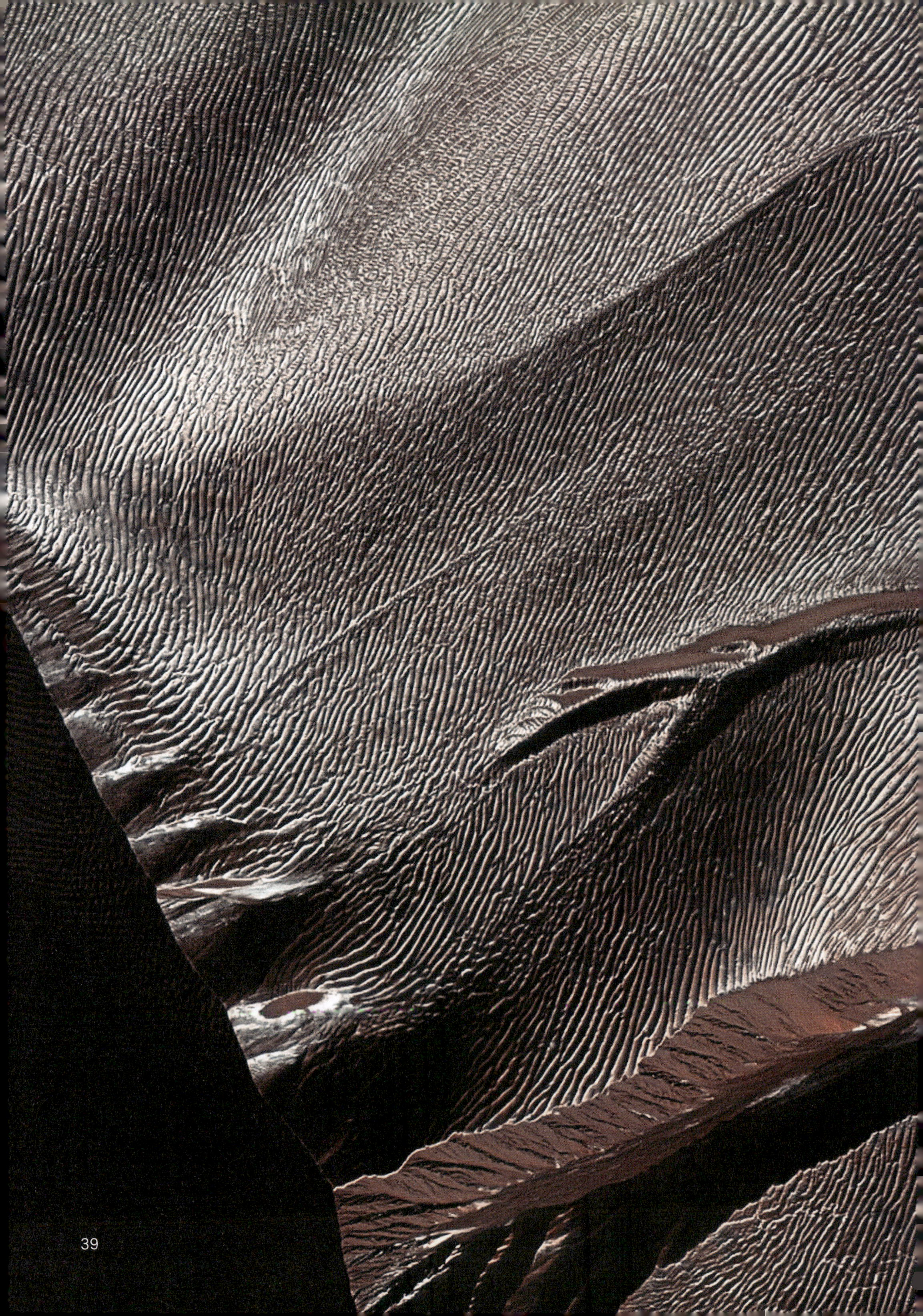

40

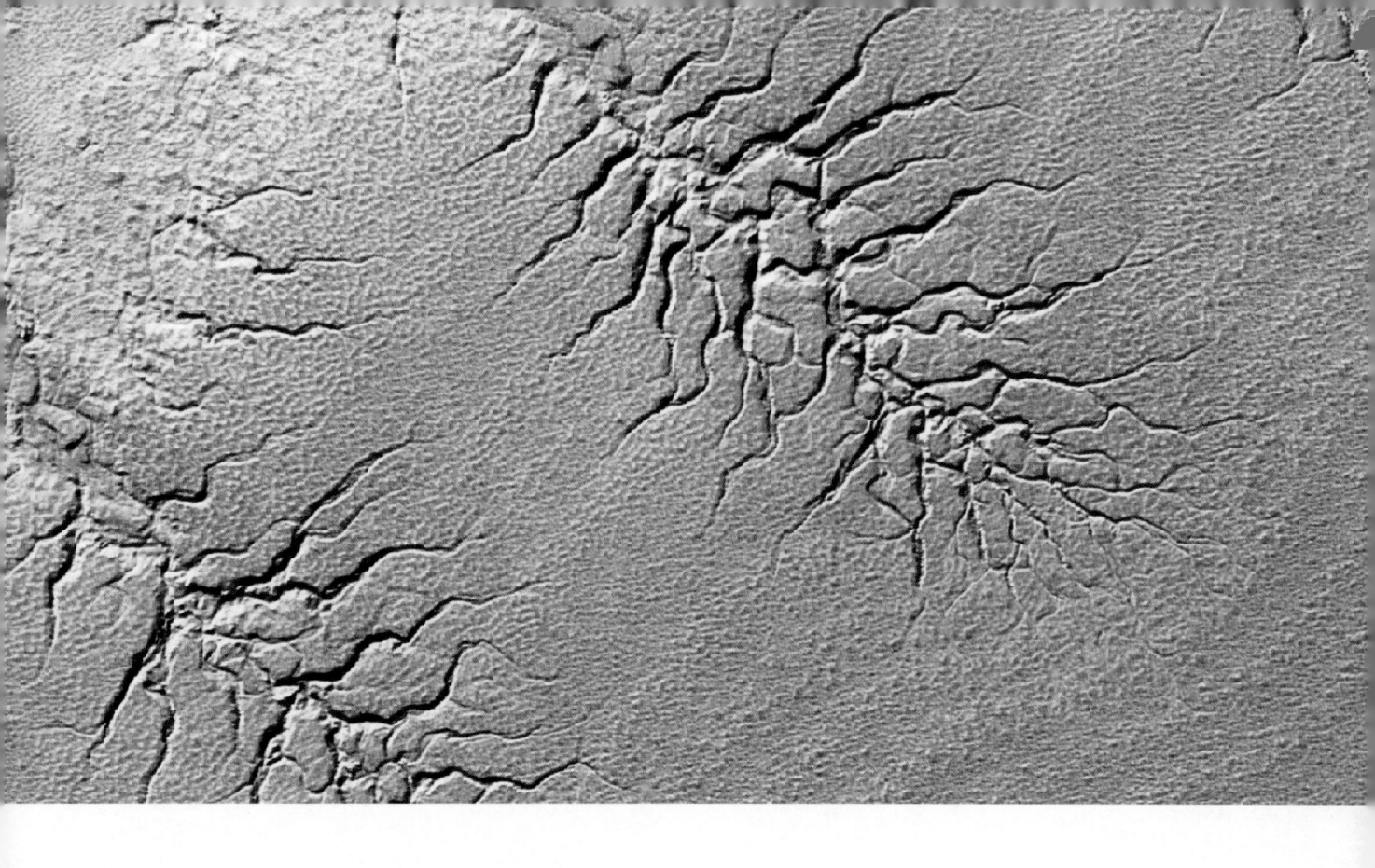

Page 41 – Image taken at the South Pole, picturing branching canals, believed to be caused by thawing carbon dioxide. These sequences are also known as Martian spiders.

Page 43 – Beautiful image of summit dunes and ripples tucked in a large sand field.

Researchers have detected signatures of hydrated salt minerals and landscapes that formed mysterious streaks on several locations on Mars. They appear during warm seasons when temperatures are above -23° Celsius and disappear in colder seasons.

Hydrated minerals have water molecules incorporated into their structure. This does not mean they are "wet" as we perceive water here on Earth but water is rather a part of their molecular composition.

The salts discovered are a mixture of magnesium perchlorate, magnesium chlorate, and sodium perchlorate. Such minerals typically form in the presence of water here on Earth and their discovery implies that a similar process could have occurred in their formation on Mars.

Nasa's rover Curiosity found these elements in the first rocks he encountered on the planet's soil. Some researchers believe that Viking 1 in the late 1970s also measured evidence of salt.

———————

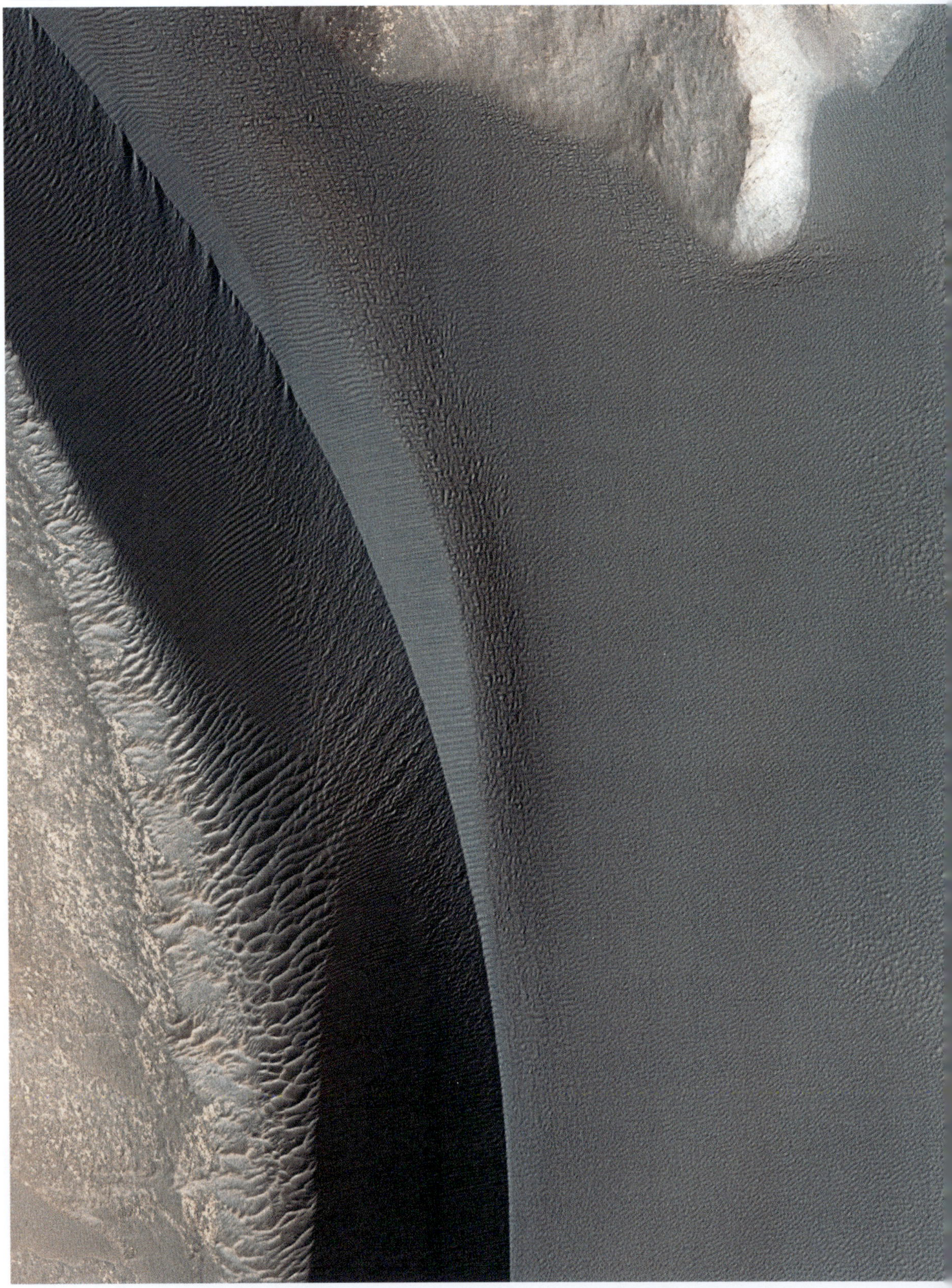

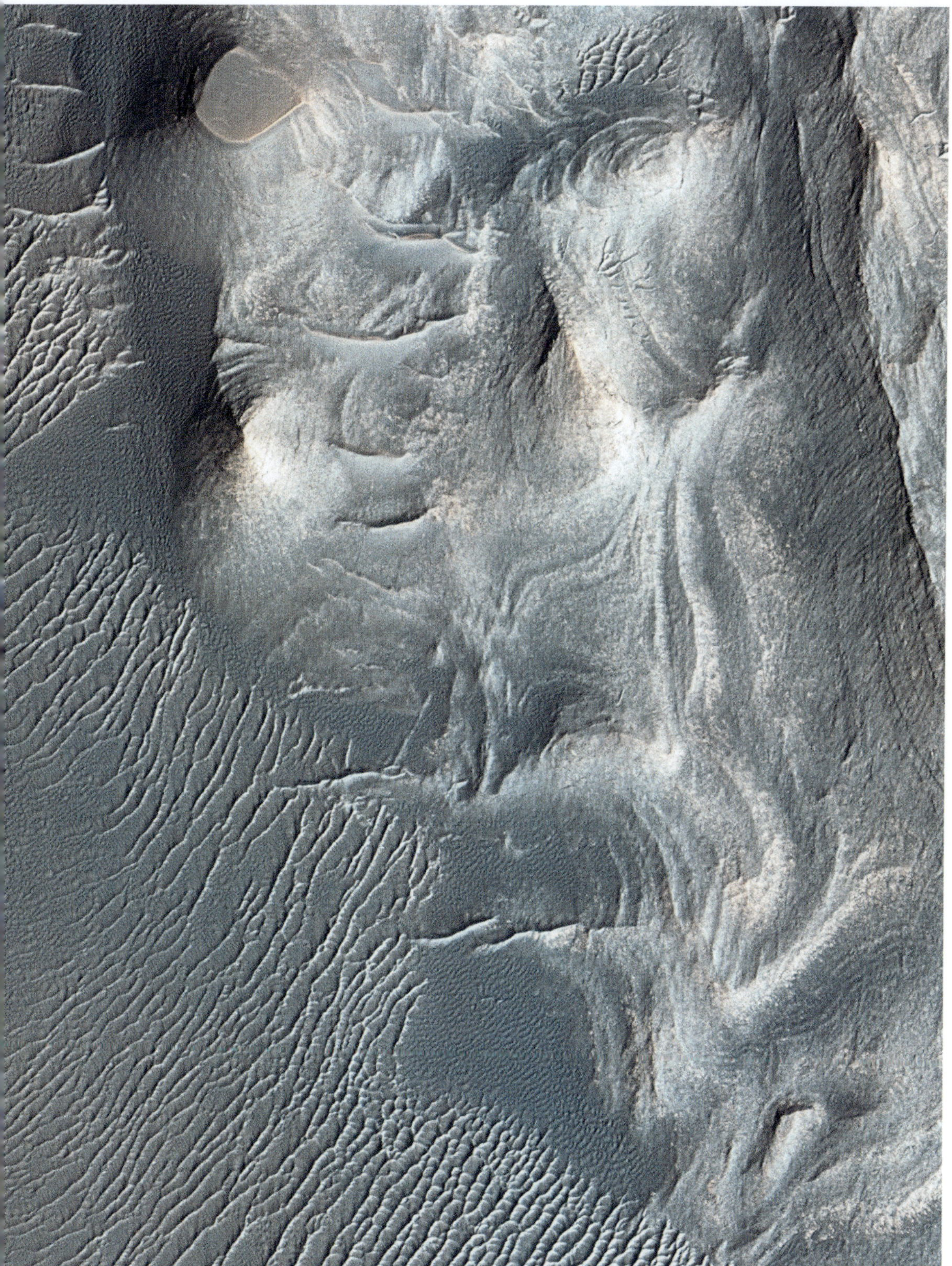

Page 46 | Sand formations in an unnamed Crater located inside the Newton Crater, captured in Terra Sirenum.
Page 47 | This image beautifully represents the splitting fissures named Cerberus Fossae – they are believed to have formed by rifts pulling the crust.
Page 50 | Beautiful display of fine-scale layering in the Gale Crater.

ENCHANTING ILLUMINATION

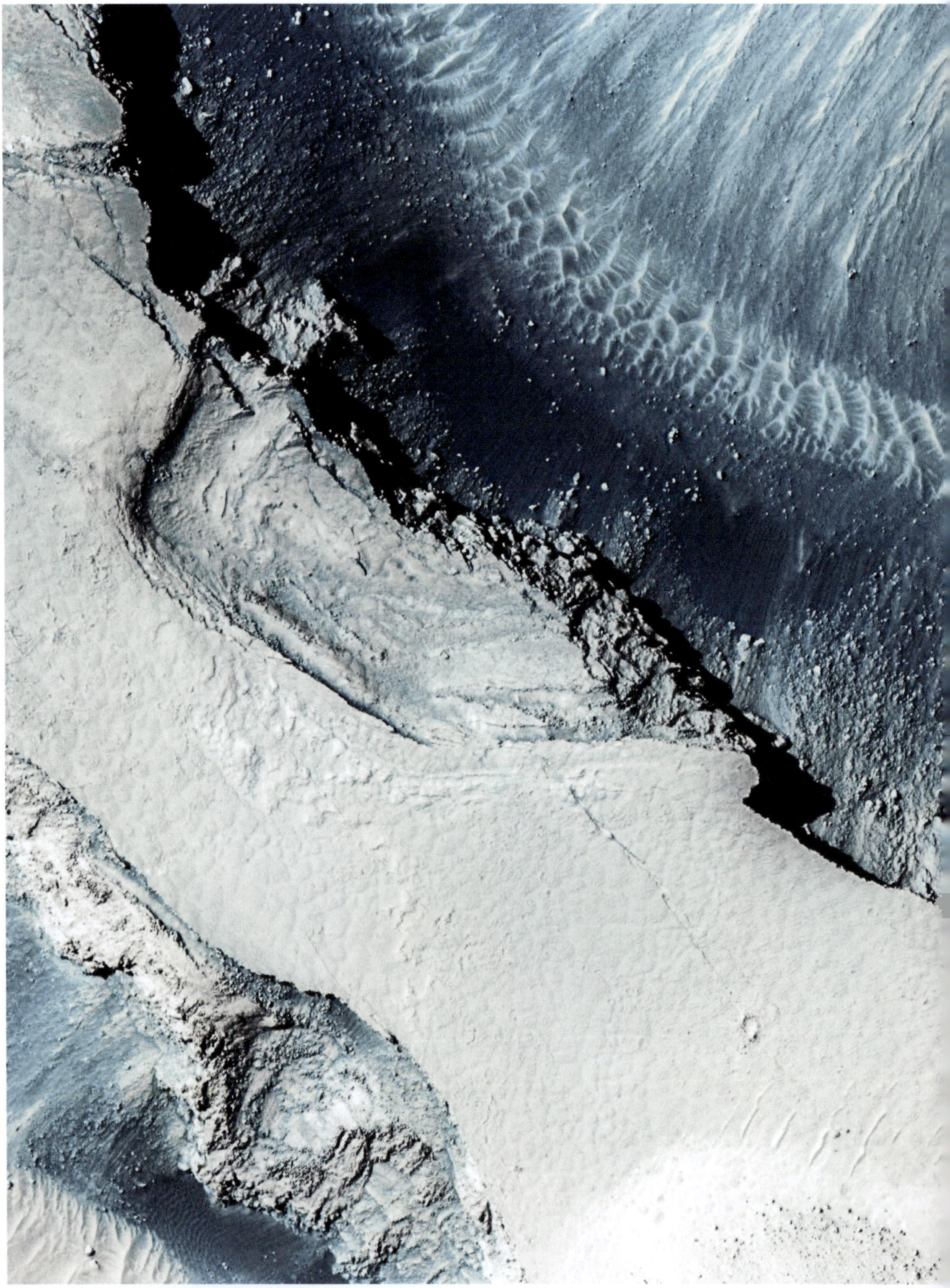

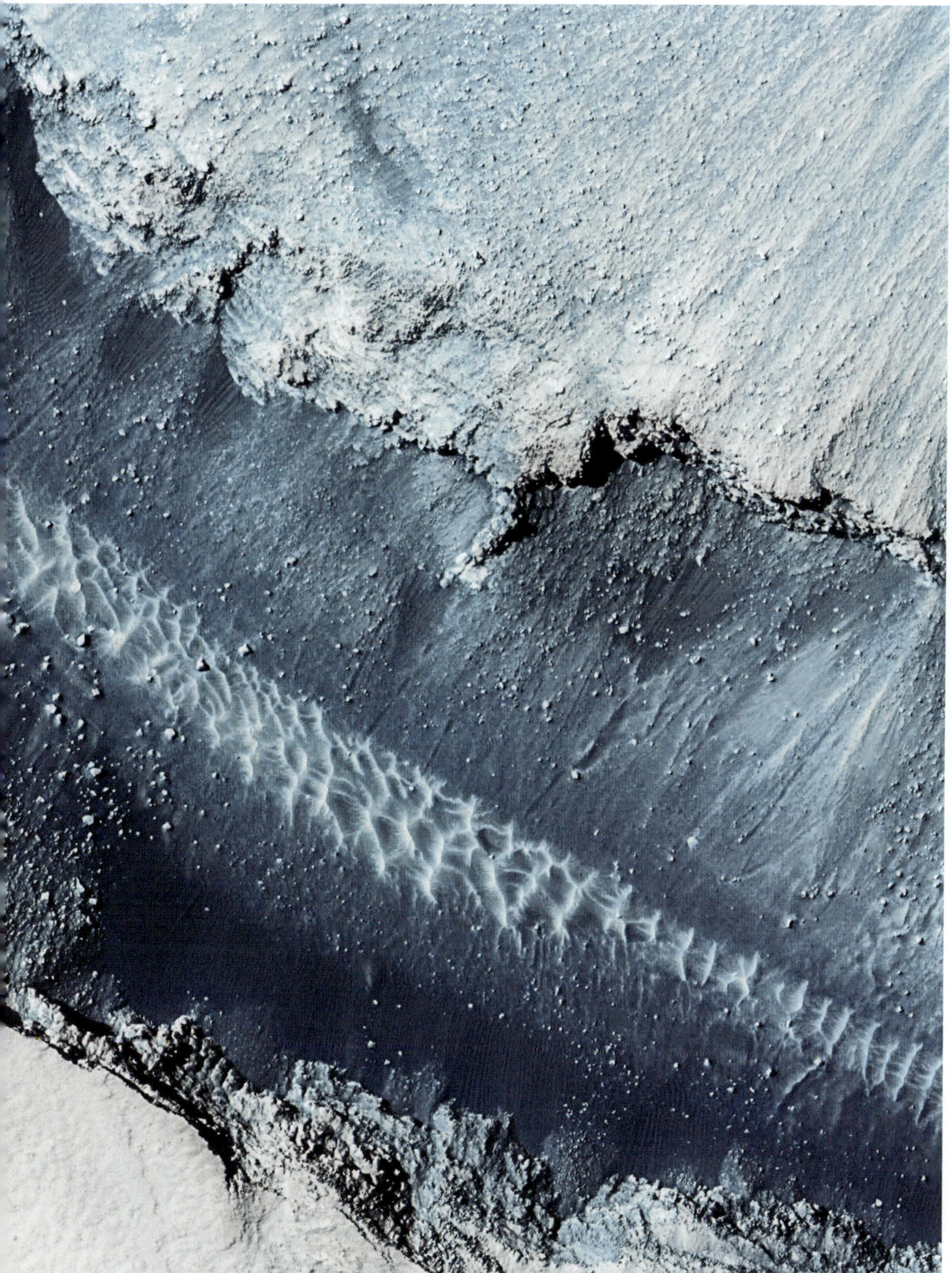

One of Mars's fascinating wonders is the phenomenon to glow in the night. This magical event was confirmed by NASA's Maven spacecraft.

A faint illumination of the Martian atmosphere glowing even without any direct sunlight is caused by a chemical reaction in its upper atmosphere.

During the day, solar radiation breaks apart carbon dioxide molecules into carbon monoxide and oxygen. At night, these atoms can form carbon dioxide again from these broken-down molecules. In this process, energy is released in the form of photons, or light, creating a mystical glow pulsating about three times per night.

Similar principles are used to create glow sticks here on earth, for fishing, or as a gadget for festive events.

The glow regularly appears after sunset on Martian evenings, creating a fascinating spectacle during the fall and winter seasons, and slowly fades towards midnight.

50

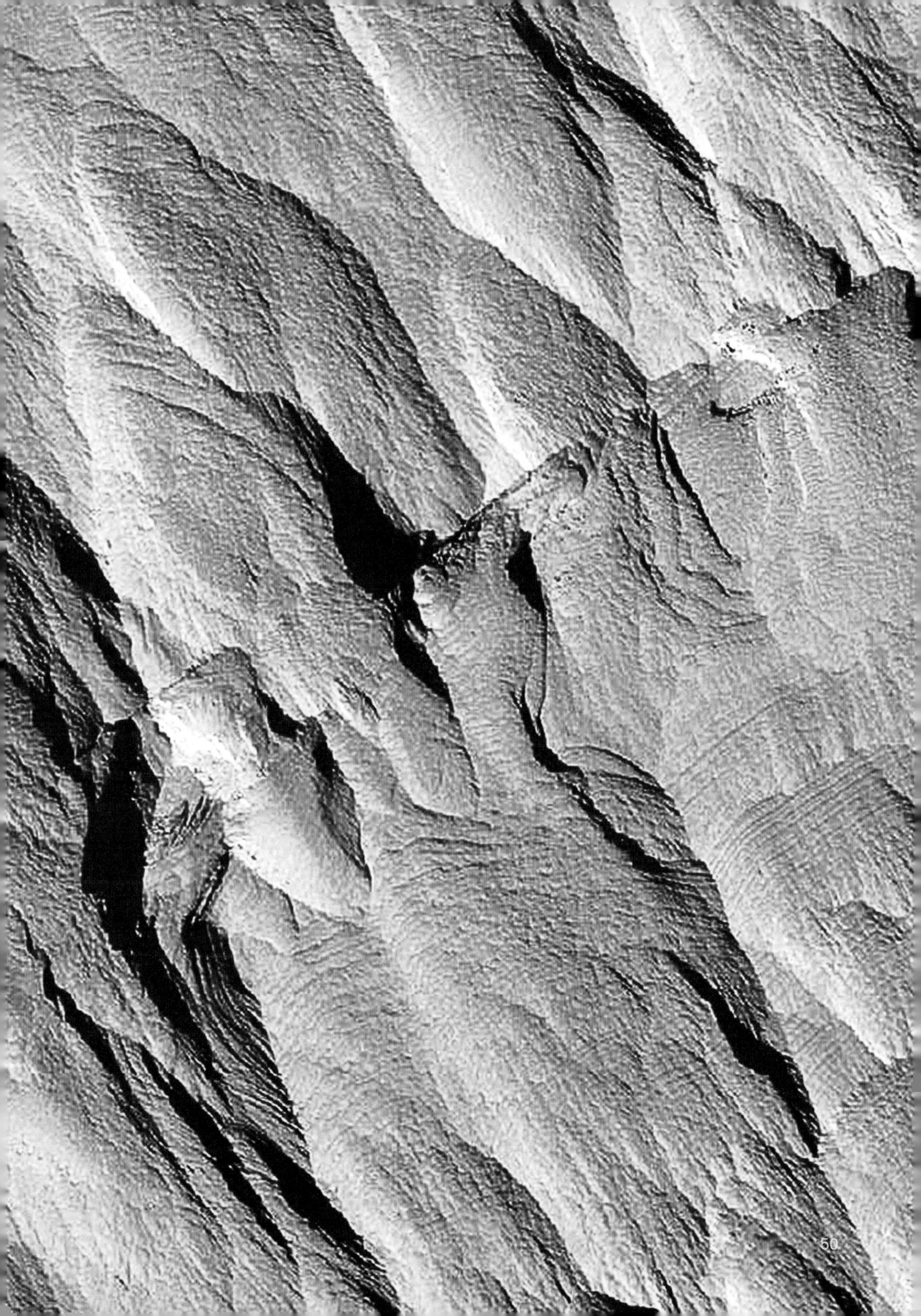

VIBRANT HUE

The perception of Mars in our heads has been red and monochromic. The planet has been associated with being red and dusty due to its surface properties absorbing the blue and green wavelength that it receives from the Sun and mainly reflecting red wavelength, giving off a rusty hue, and rising to a 'red' planet to our eyes.

Contributing to the appearance of its signature red shade, most rocks contain a lot of iron that oxidizes and turn reddish. However, up close, the planet's colors are more of a beige hue, and some landscapes are golden, tan, or even green-toned in certain areas.

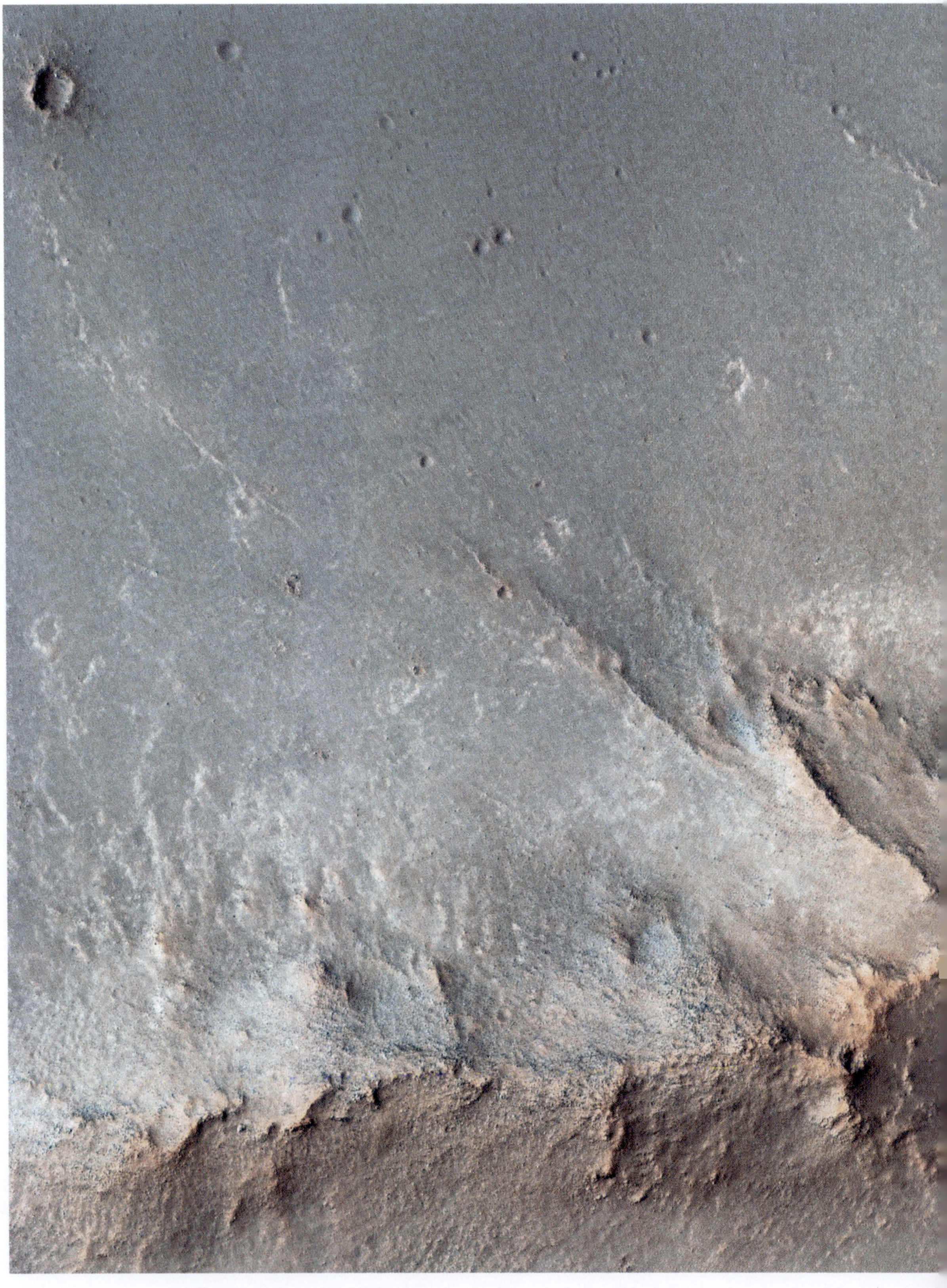

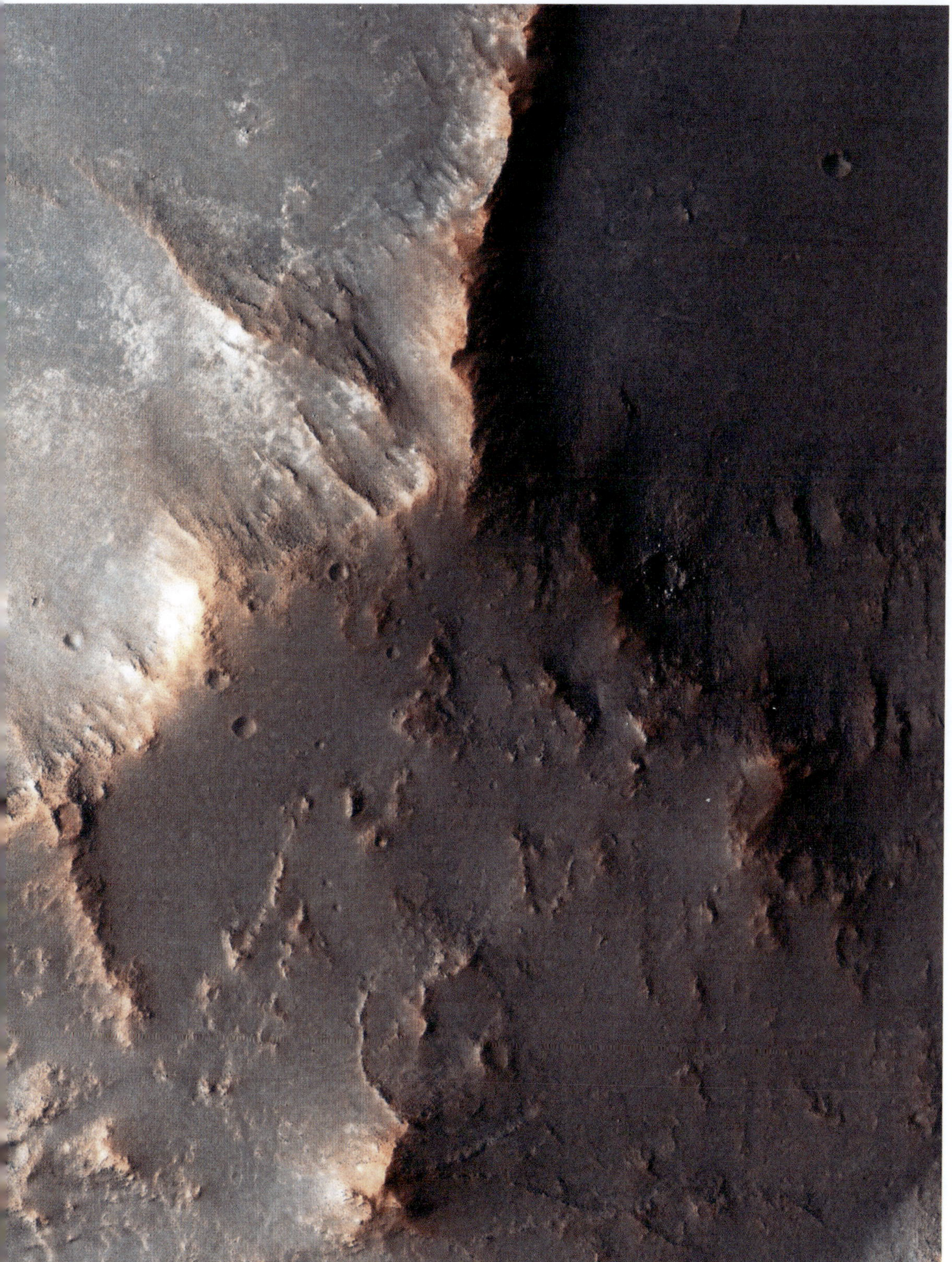

CELESTIAL ALCHEMY

Page 51 | Gorgeous formations inside an impact crater located in Isidis Planitia.
Page 53 | The Compact Reconnaissance Imaging Spectrometer (CRISM) has detected the presence of aluminium, iron, and magnesium in this area. Such discoveries are important to understand how such rocks form, adding to our understanding of Mars.
Page 55 | The largest canyon system in our entire solar system is Valles Marineris. In this image, we see a capture of Ius Chasma, a significant part of the several Canyons contributing to the largest canyon system.

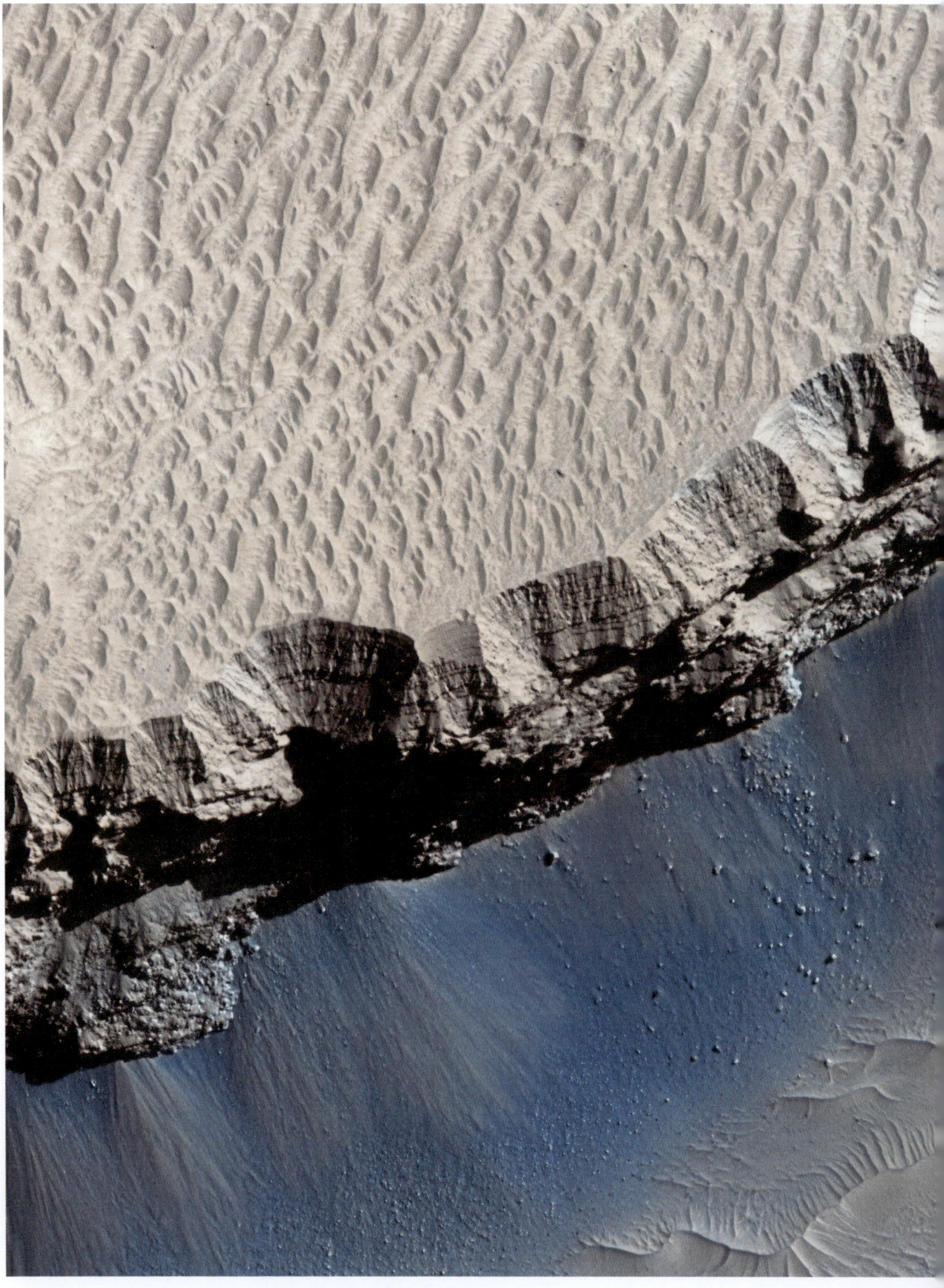

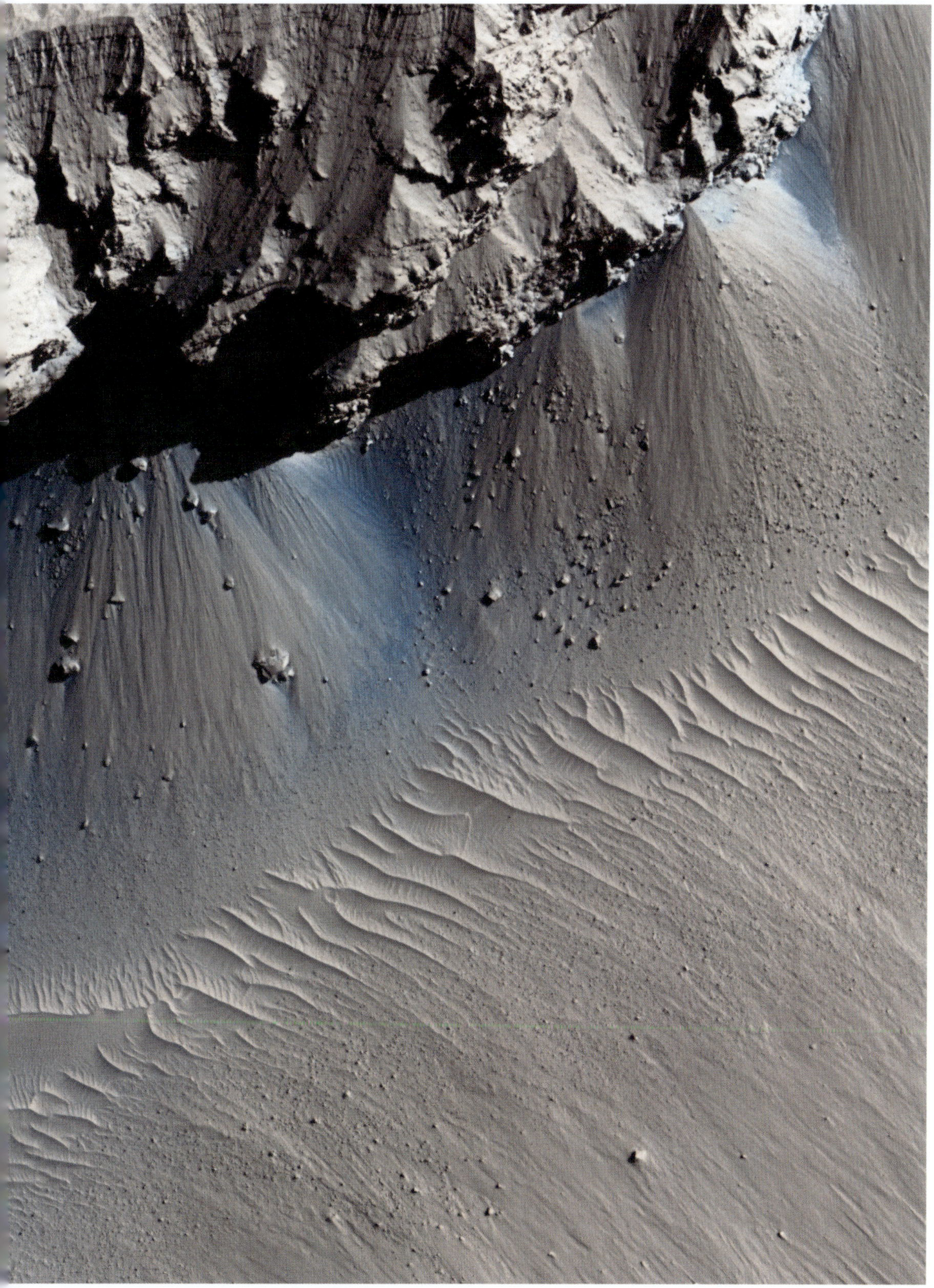

Page 57 | Image captured in the region of Cerberus Fossae. Such formations are believed to have been formed by Marsquakes in the past.
Page 61 | Beautiful display of deposits located in the Gasa Crater, captured at springtime of the planet. However, it is believed that these deposits have formed during Mars's winter and are caused by seasonal carbon dioxide frost.

Mars's atmosphere has captivated scientists with its complex arrangements and insights into the planet's history, the potential of past life, and its habitability.

The atmosphere on Mars is thin and roughly 95% carbon dioxide, 3% nitrogen, 1.6% argon, and traces of oxygen and other gases. To give it a perspective, our planet Earth has 78% nitrogen, 21% oxygen, 1% Argon, and only 0.04% carbon dioxide along with small amounts of other gases.

Mars only holds 0.13% oxygen in its atmosphere, compared to 21% on Earth.
If we ever want to visit Mars, we either have to bring it along or make it ourselves.

Page 63 | Dunes on Mars are classified by their shape. These gorgeous dunes are called star dunes have a central peak and three or more arms. Such shapes get formed by winds from different directions.
Page 65 | Spectacular image showcasing exposed layers in Noctis Labyrinthus.

61

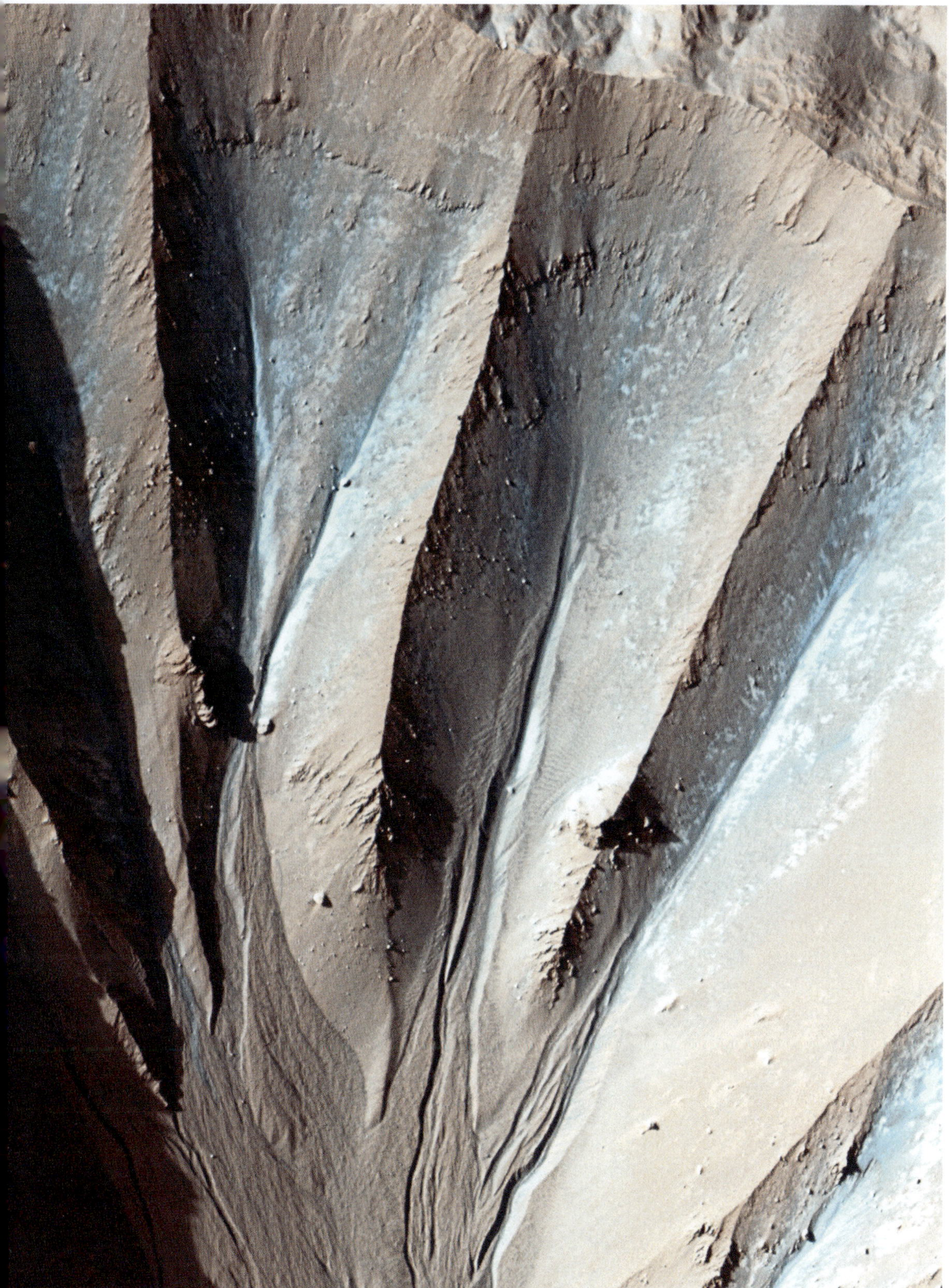

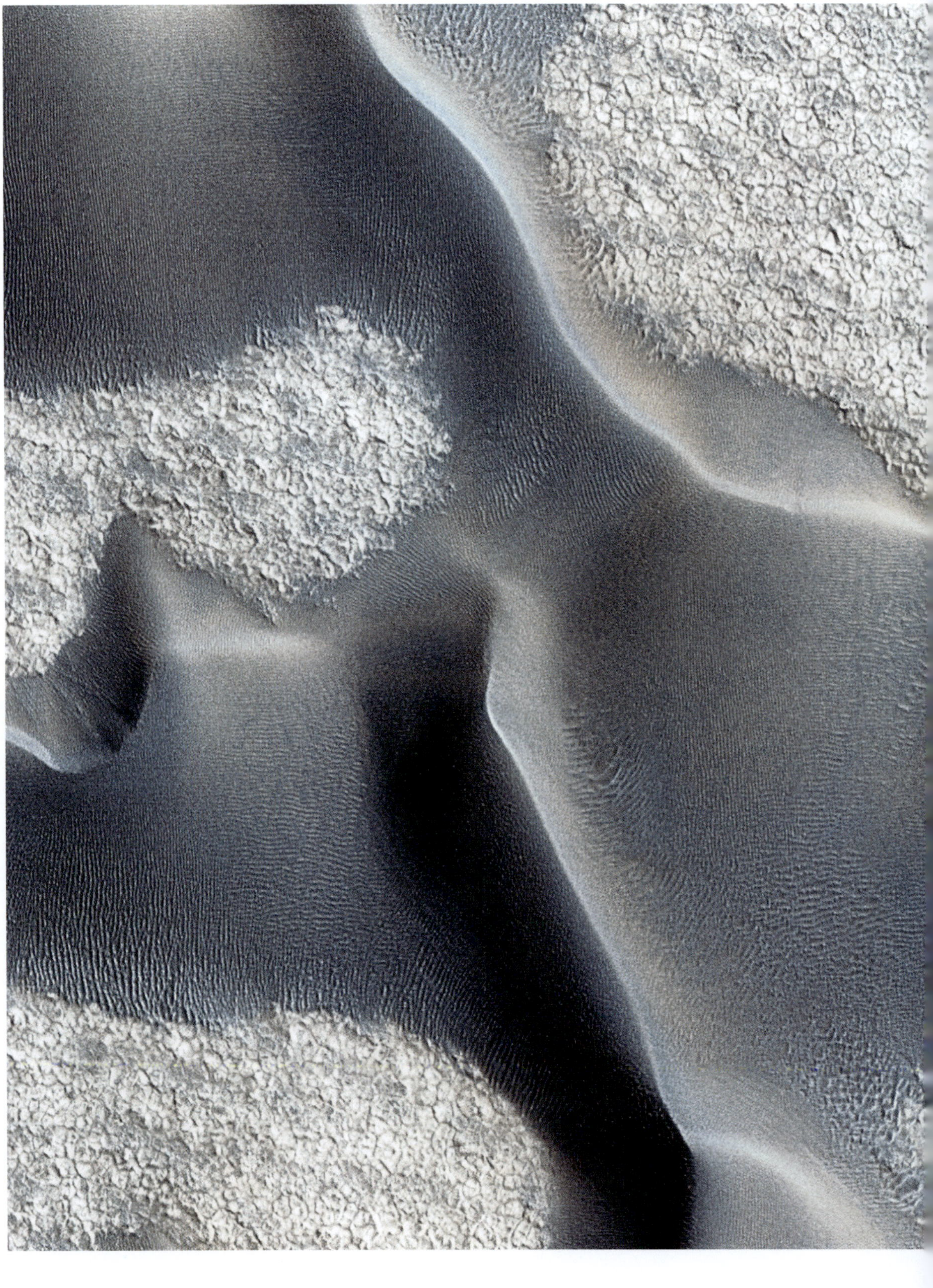

64

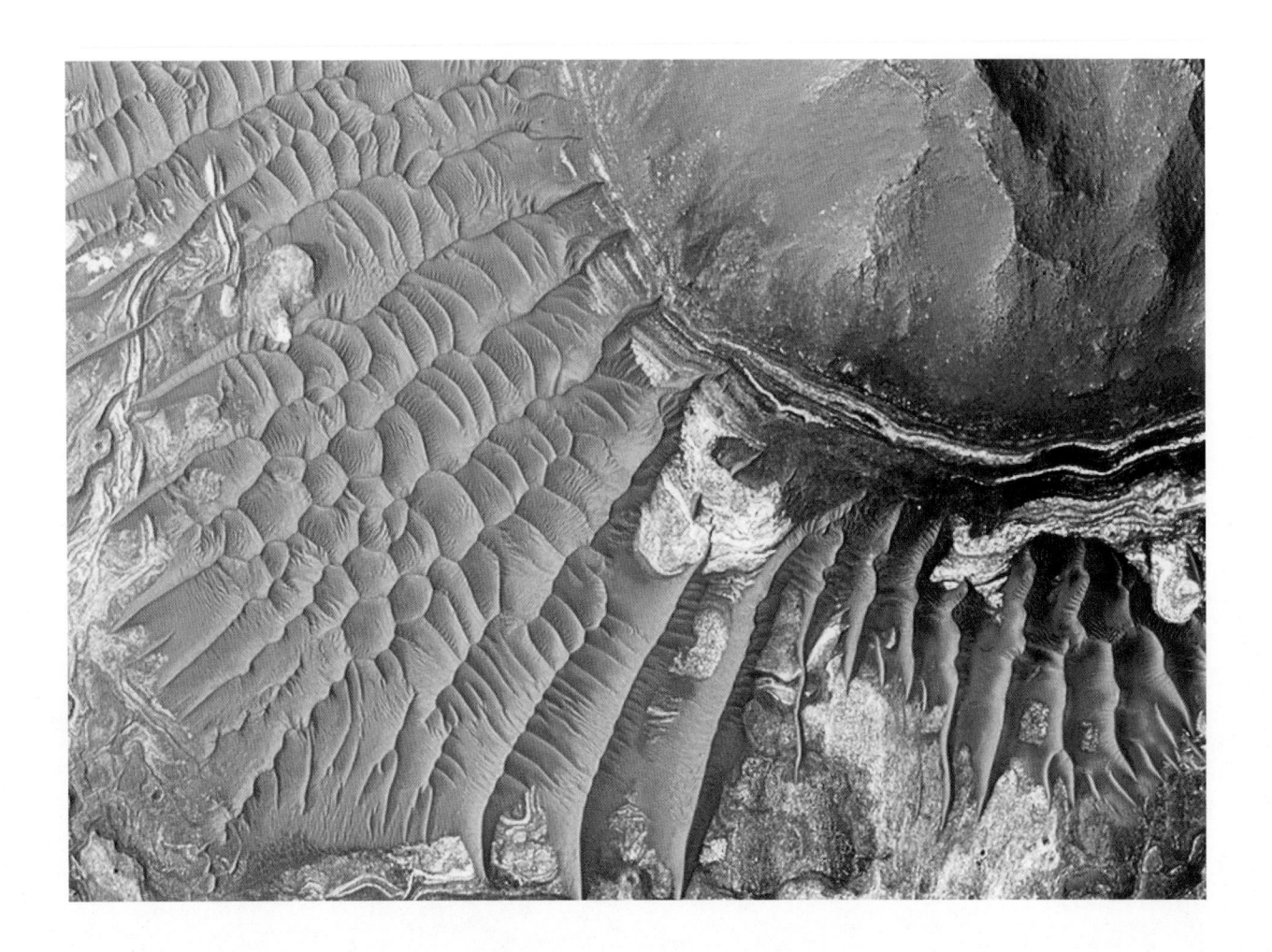

Mars's atmosphere, despite being thin, plays a crucial role in the Martian weather, such as temperature variations and dust storms, creating the primary feature on Mars, the Dunes.

The planet's panoramas are breathtaking and enchanting. Some formations present themselves in artistic and captivating ways, leaving its mankind viewers fascinated.

Mars's dunes are formed by Martian winds that only have a force of 1% compared with winds of the same speed on Earth, a consequence due to its thin atmosphere.

The impact is still potent enough to raise small and fine dust particles, enough to form wispy clouds and dust storms.

This power is mainly created due to solar radiation that heats the planet's atmosphere. In addition, Marst has a minimal surface gravity of 3.711 m/s2, making it easy to lift dust particles. In perspective, Earth has a surface gravity of 9.807 m/s2. This means that the speed of an object falling freely will increase by about 9.8 meters (32 ft) every second.

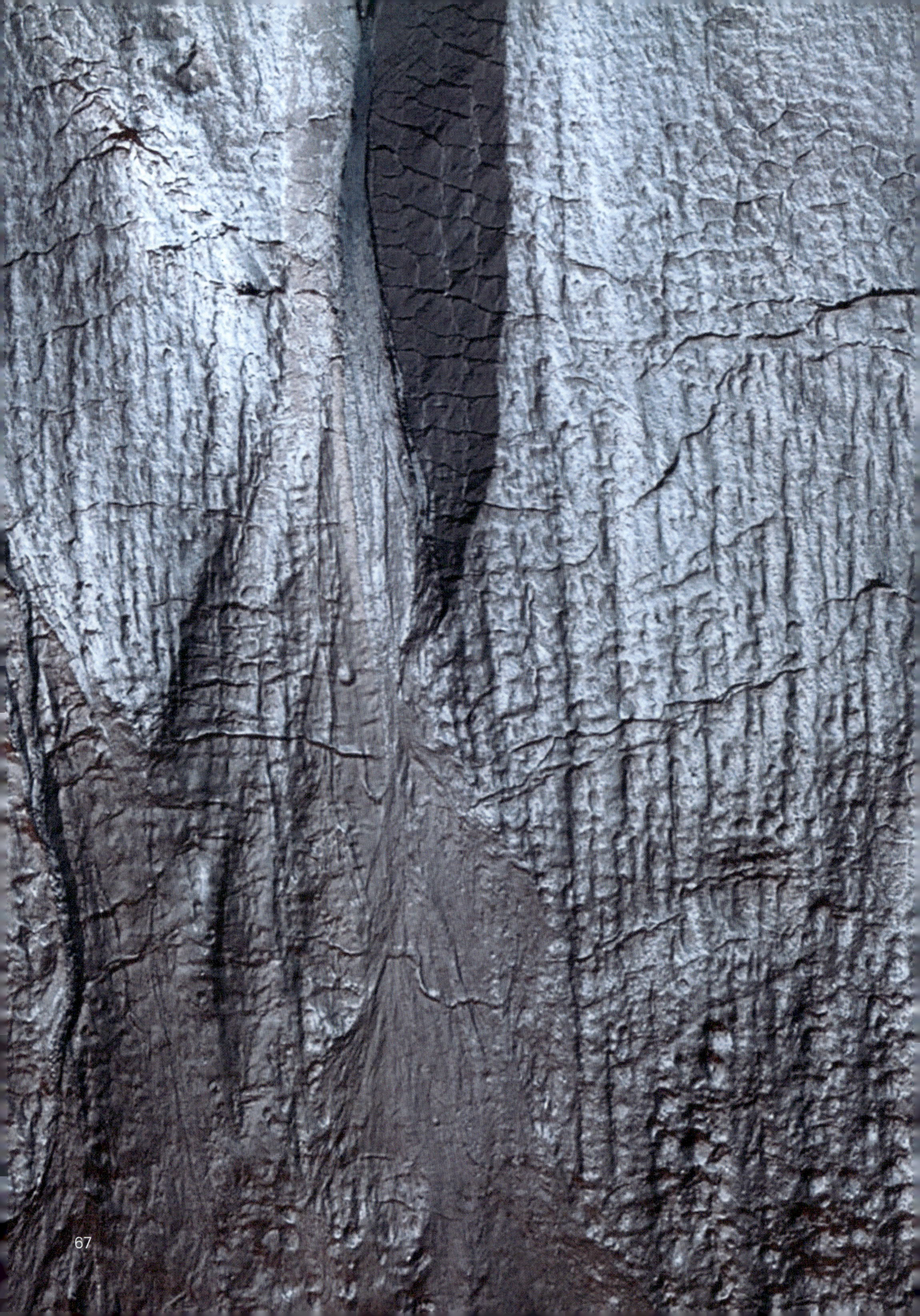

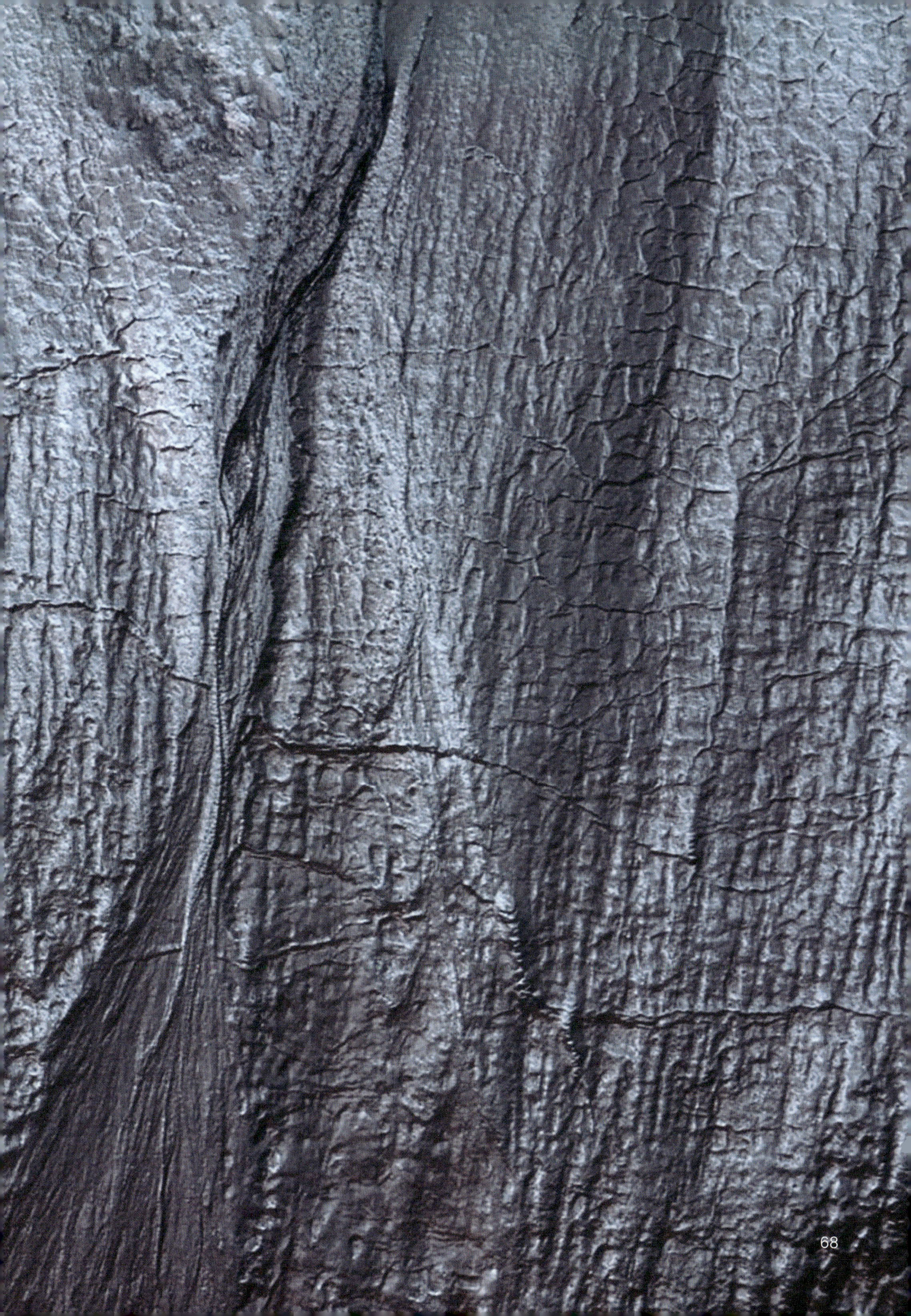

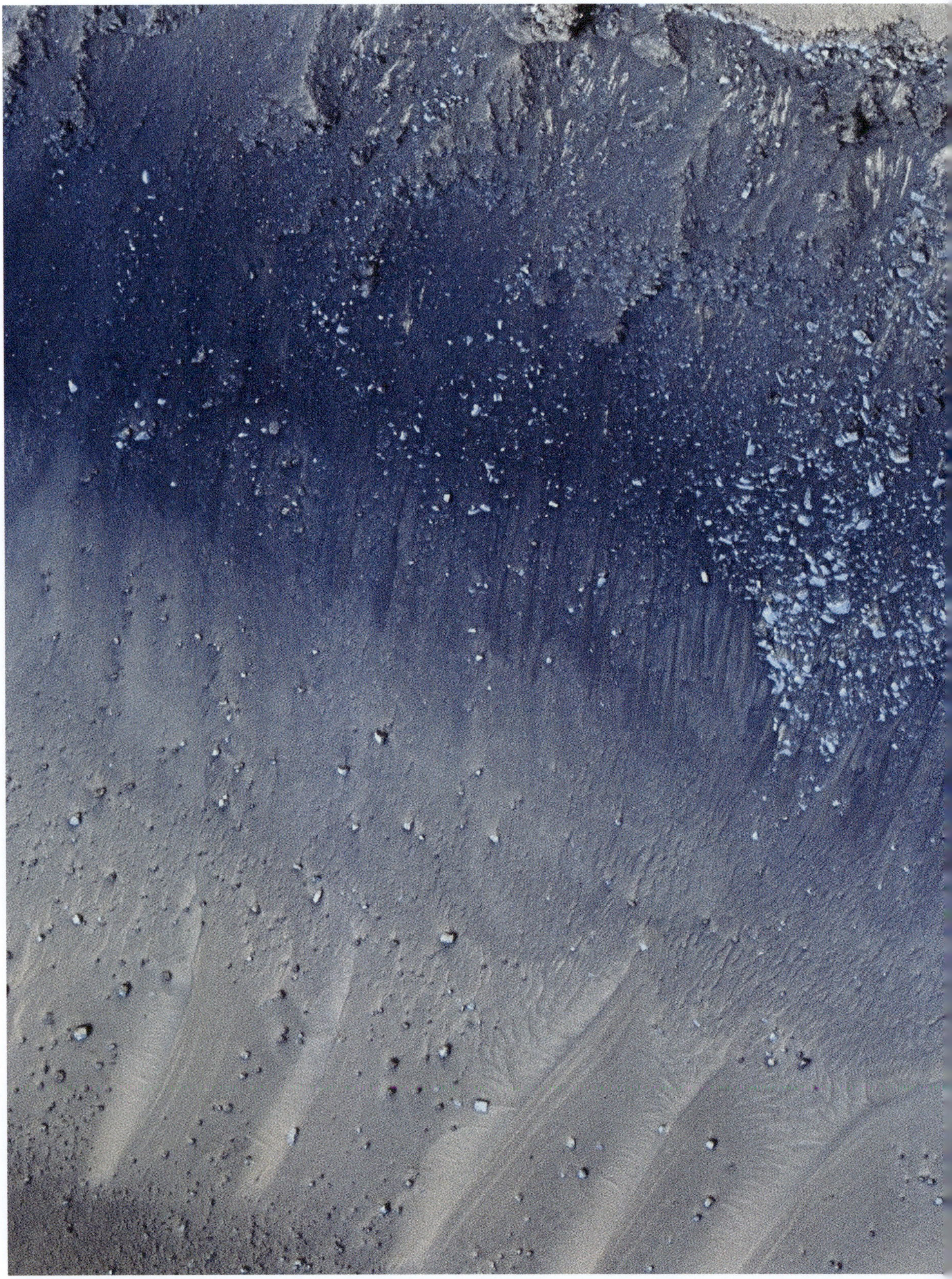

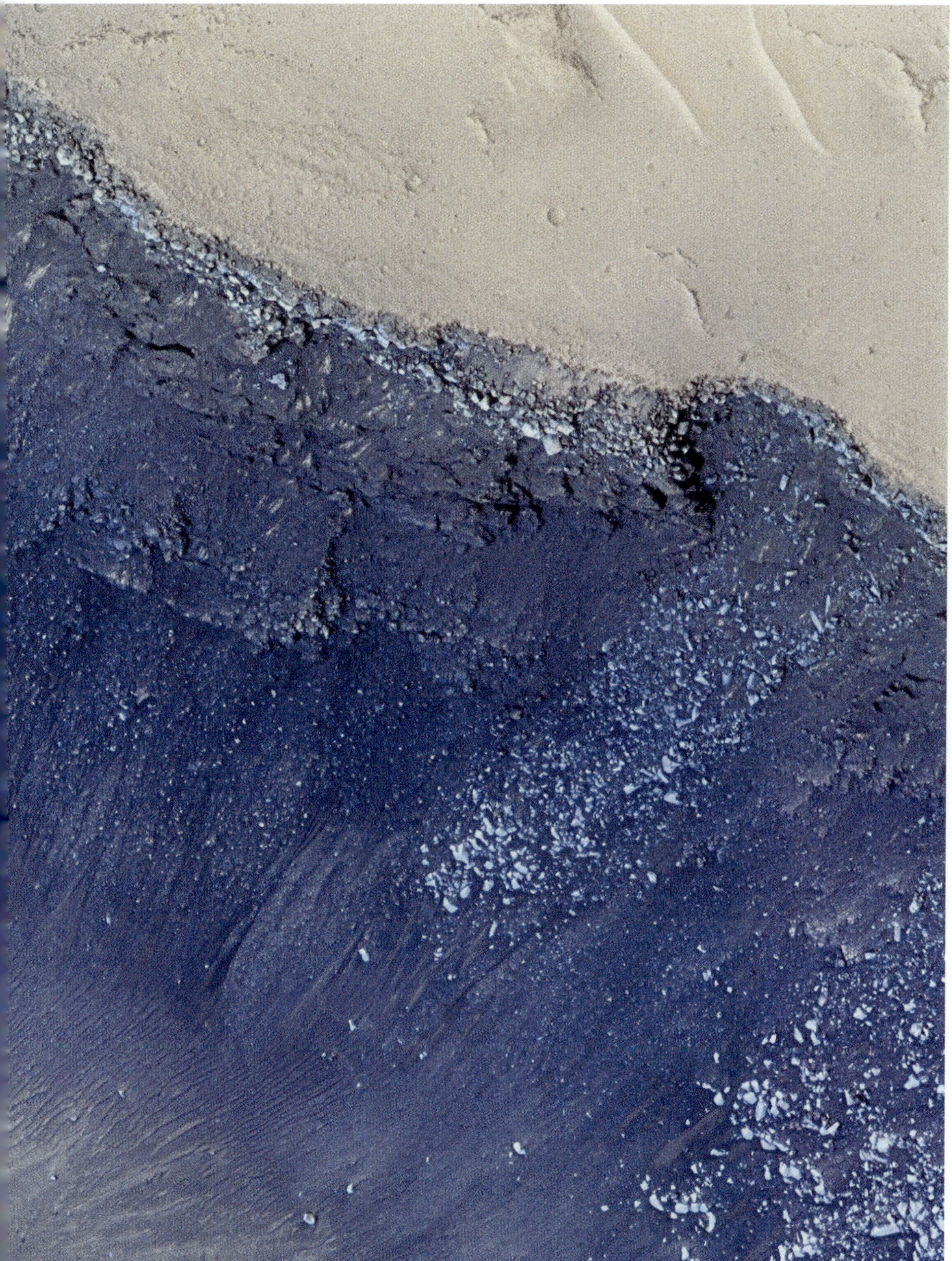

It is believed that Mars's atmosphere has been much thicker in the past, which could've supported liquid water on the surface.
Over billions of years, the atmosphere gradually degenerated, driven by the solar wind streams of charged particles from the Sun.

Without a global magnetic field to protect Mars's atmosphere. The Planet was vulnerable to the Sun's winds.

Page 67 | Image displaying beautifully how gullies form during the plant's winter trou fluids of carbon dioxide frost.
Page 69 | Active lands on steep-sided parts, named Cerberus Fossae, on volcanic plains at Elysium Mons.
Page 72 | Beautifully deposited sand of shallow channels. Mars's thin atmosphere is not able to shatter light much, which displays shadows sharp and dark on the planet.
Page 73 | Extraordinary dunes made of a specific type of sulfate mineral, named gypsum, and captured at the Martian north pole in Olympia Undae.
Page 75 | The Majestic Victoria Crater is about 800m (2,600 ft) in diameter and located in the Meridiani Planum, near the equator of Mars.

MYSTIC PATTERNS

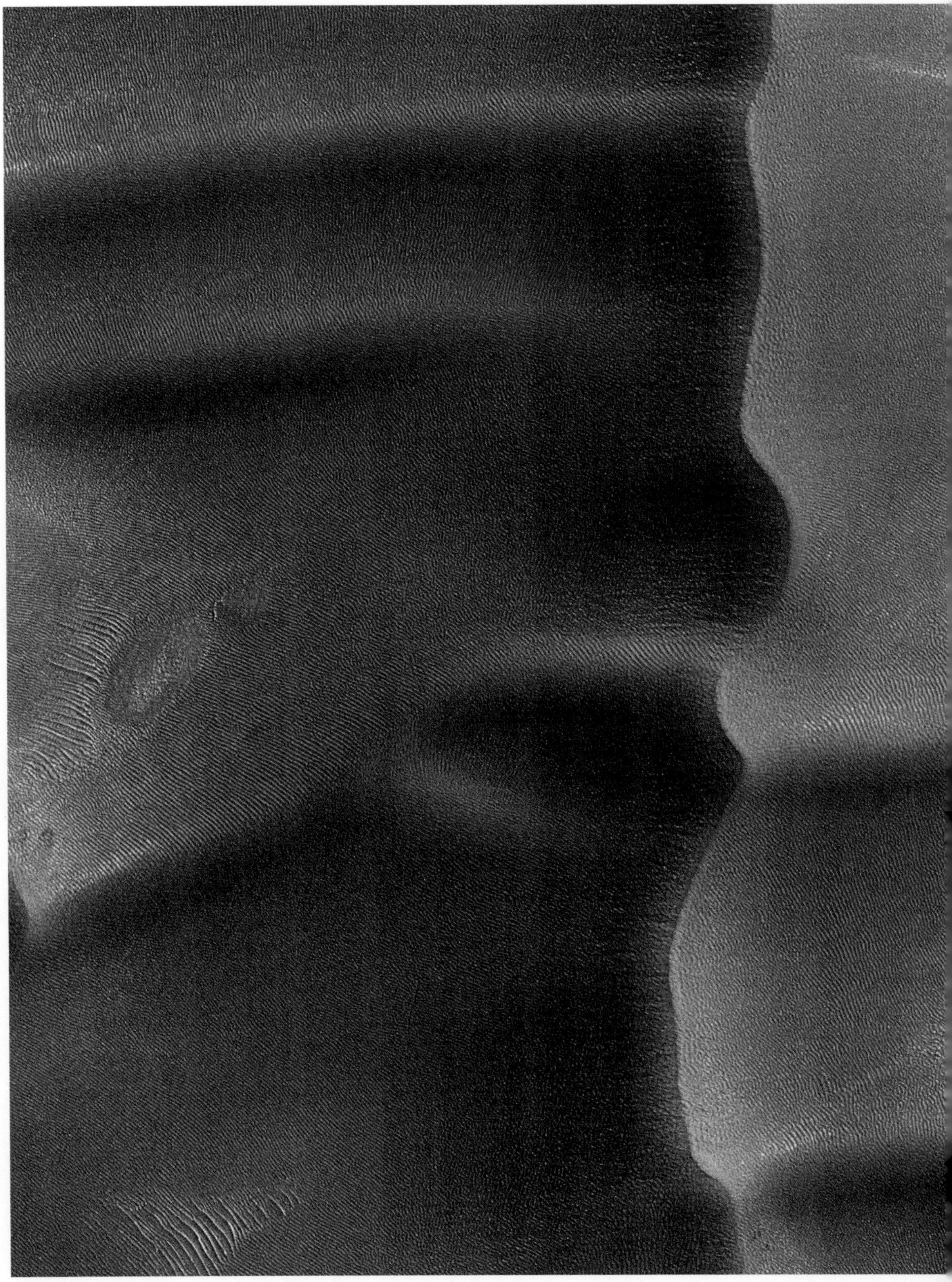

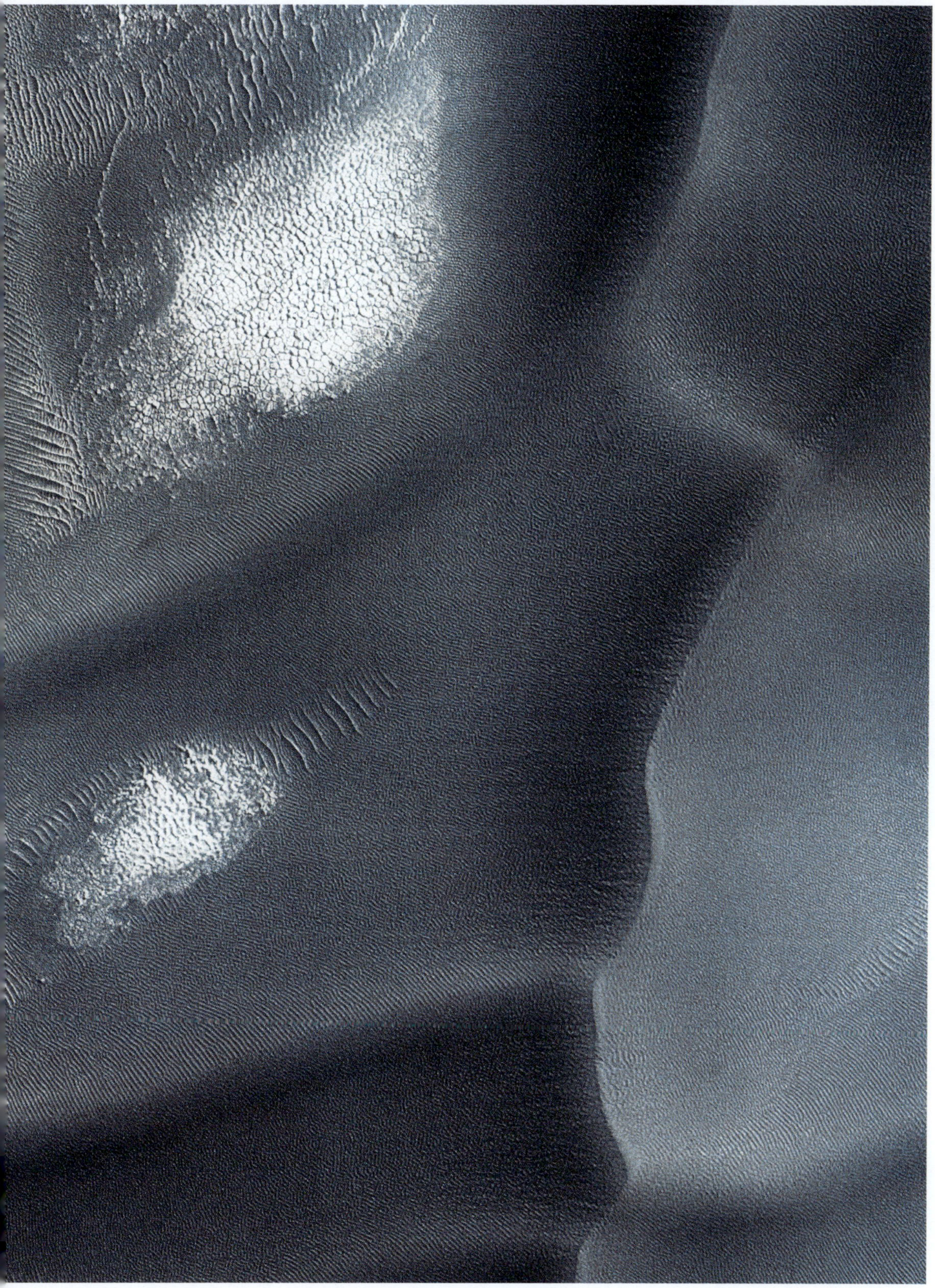

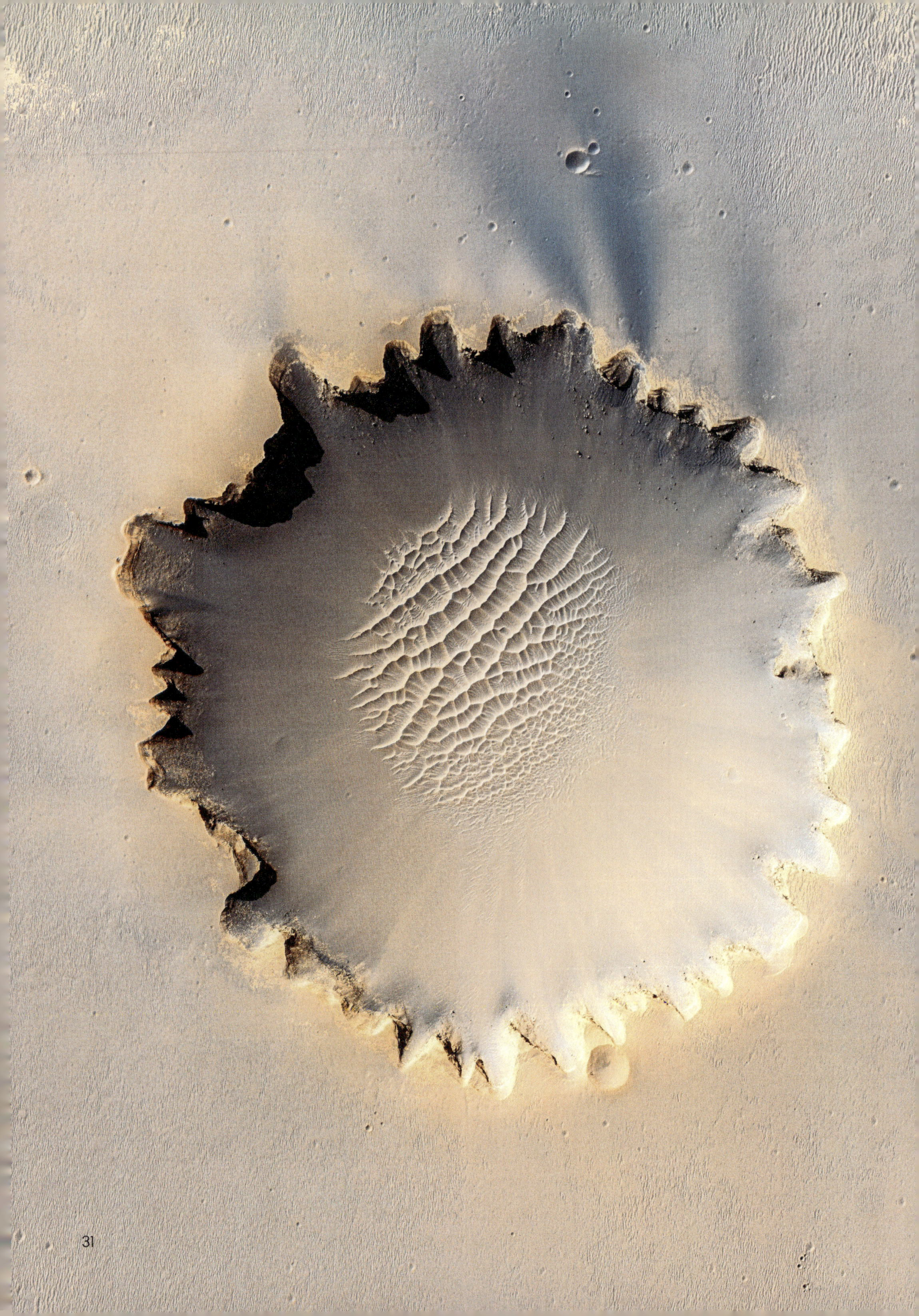

MAJESTIC IMPACT

Craters on Mars exist by the hundreds of thousands, but only creators that could potentially hold significant interest for research are given names - applying to around 1,000 craters.

Some hypotheses suggest that the entire northern hemisphere on Mars is a massive impact basin named the Borealis basin. This theory could explain why the northern hemisphere of Mars is a consistently flat plain and, on average, 5 km (3.1 mi) lower than the entire southern hemisphere. Another factor supporting these speculations is the fact that the crust is thinner in this region compared to others.

77

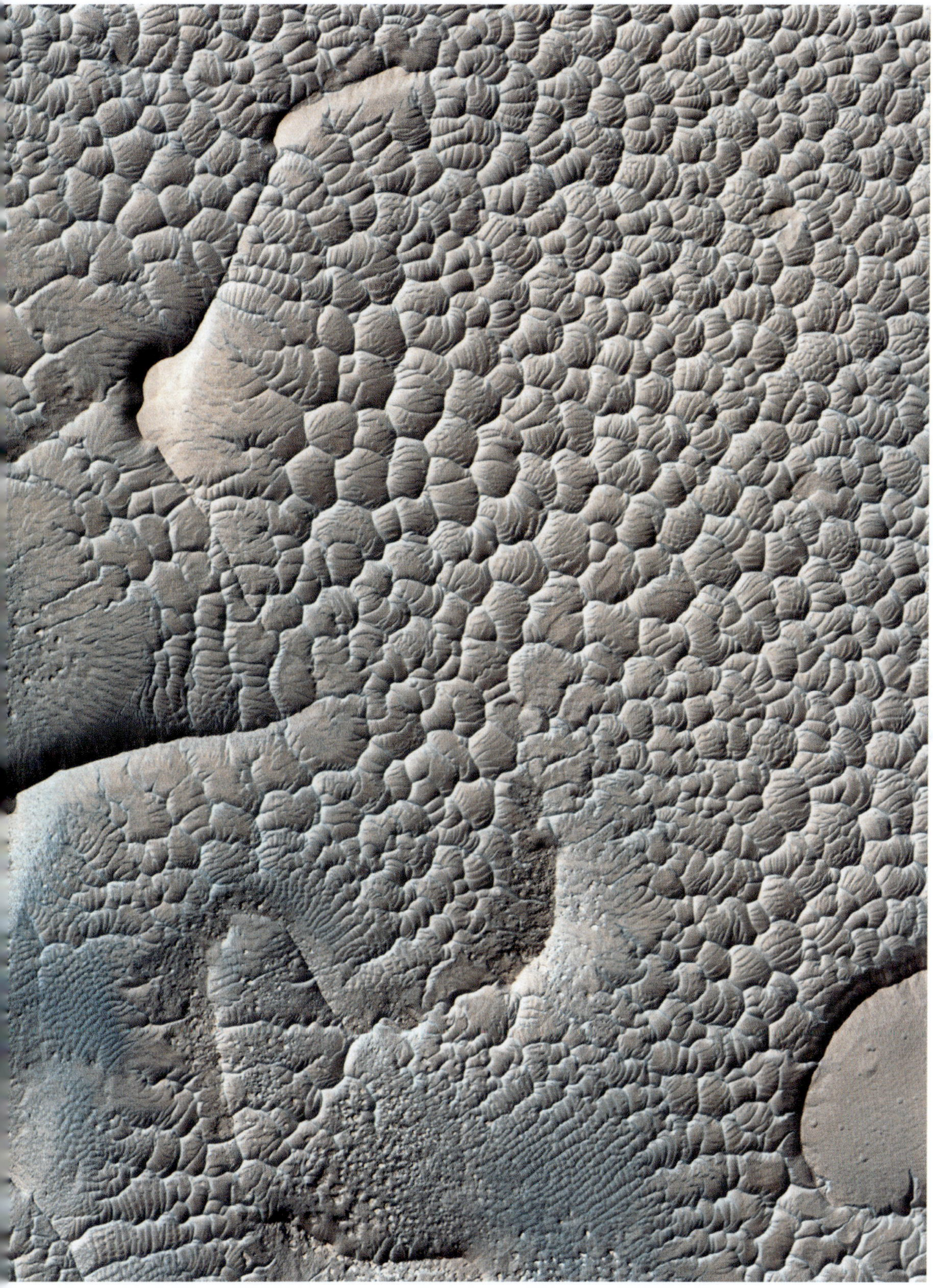

Hellas is the largest recognized impact crater in our entire solar system, blasted by an enormous asteroid and 2,300 km (1,400 miles) in diameter, and a magnificent 9km (5 miles) deep.

Hellas impact basin is often mentioned as being the lowest point on Mars, but the exact point is a much younger and smaller impact crater, although itself located inside of the Hellas crater.

Page 77 | Gorgeous dunes turned into stone and frozen in time. The formation is located in south Melas Chasma.
Page 80 | Land covered in ice sheets in Arabia Terra. Cracks along the spines are believed to have formed due to the contraction and expansion of the ice.

ARABIA TERRA

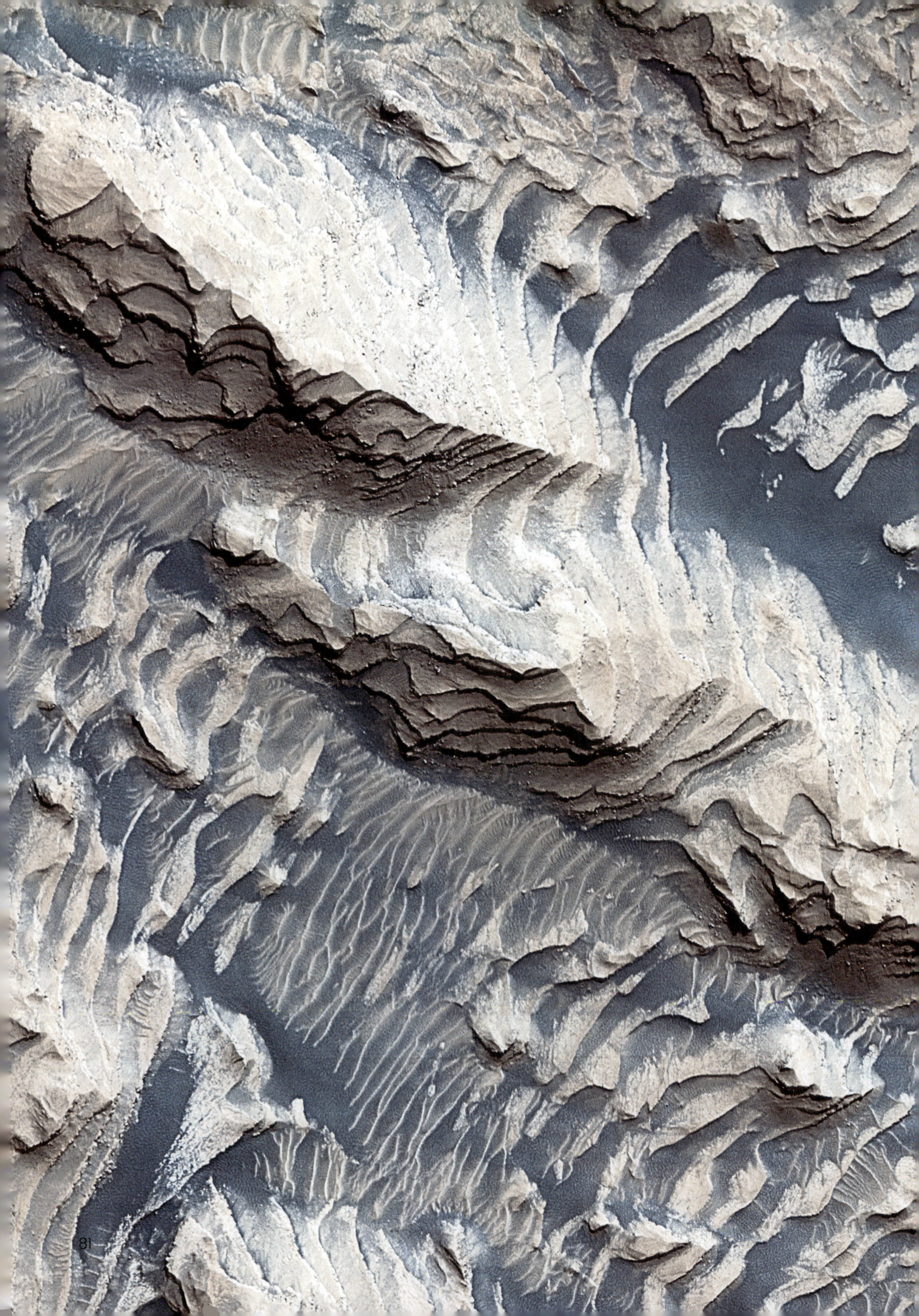

Page 81 | Sedimentary rocks in Arabia Terra, formed millions or even billions of years ago, with layers of loose sediment cemented into place, forming stunning arrangements
Page 83 | Rocks covered in dunes in the western Nereidum Montes, composing an enchanting formation.
Page 85 | A closeup of alluvial fans in the Saheki Crater. The area around the crater has some of the best-preserved alluvial fans on the planet.

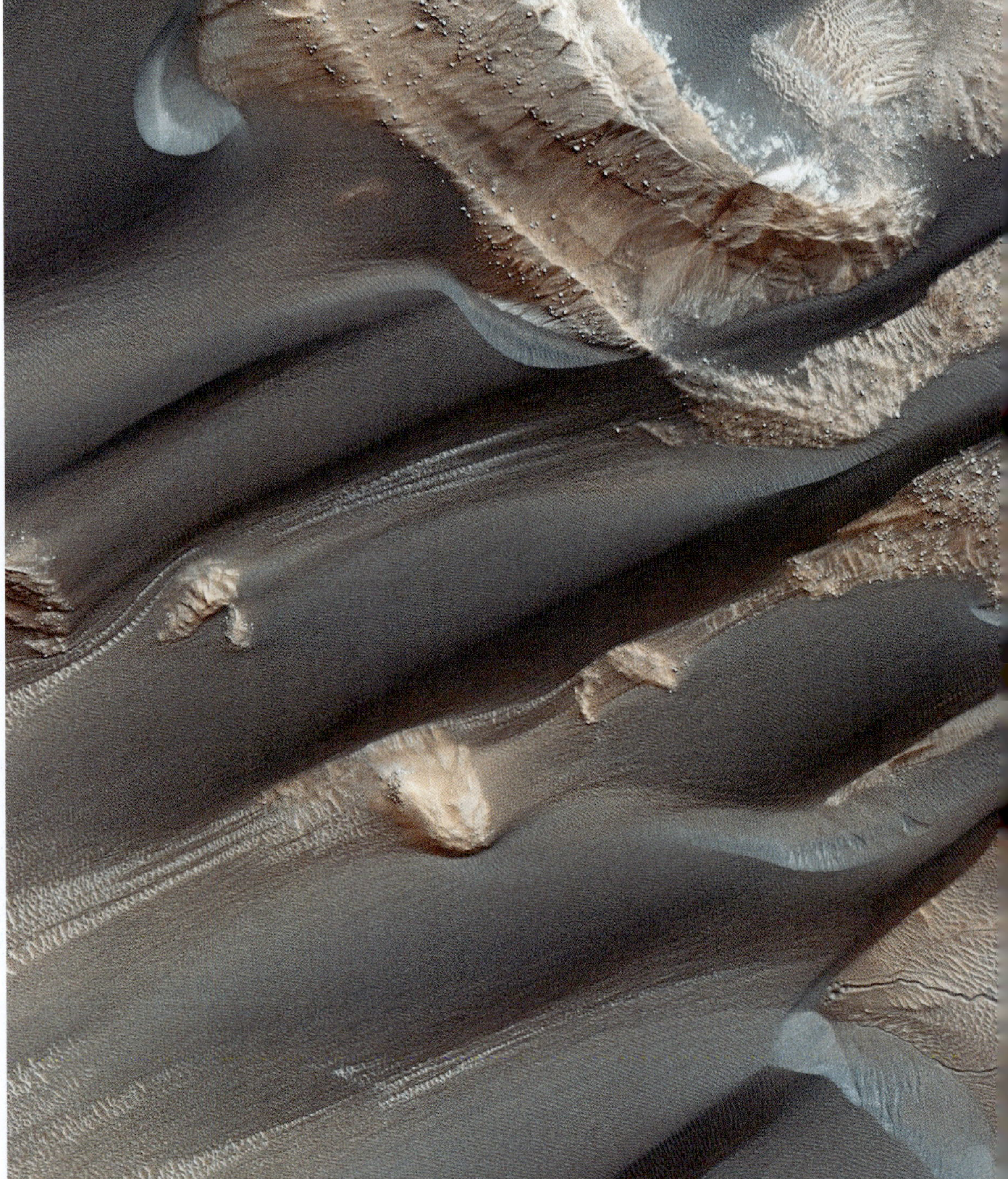

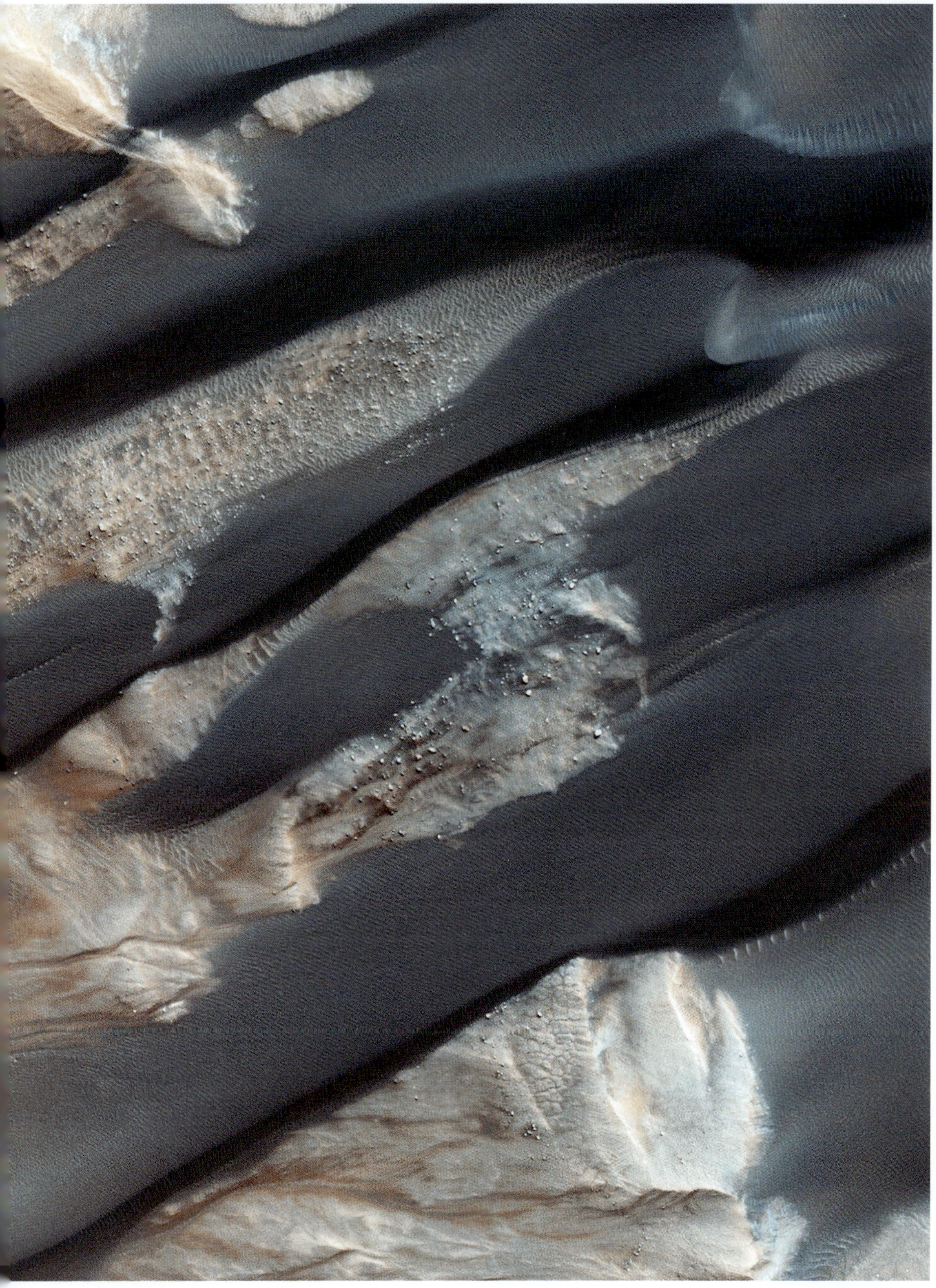

Arabia Terra is the largest Terra on Mars, densely cratered and heavily eroded, with a length of 600 km at its greatest extent. The land was named in 1876 by Giovanni Schiaparelli after the Arabian Peninsula, the largest peninsula on Earth. Peninsulas are called large landscapes, surrounded by water but connected to the mainland on one part.

The Highlands of Arabia Terra are located in the north and are speculated to be one of the oldest regions on Mars. It is presumed to have a watery origin, estimated to have formed around 4 billion years ago.

Its land is rough and densely cratered, featuring a lot of roll-down slopes. It is believed this could've been the result of lava having once flowed through this region.

Other regions in Arabia Terra, feature layered rocks with alternating light and dark material, these may be the result of episodic periods of ancient lakes and changing shorelines, as the sea level has varied with the Martian climate.

———————

Page 88 | Stunningly elongated paint inside an impact crater located in the southern polar region on Mars.

Page 89 | An impressive collapsed ceiling of a lava tube north of Pavions Mons. The skylight is approximately 35 m (115 ft) wide, and the cave is estimated to be around 20 m (65 ft) deep.

Page 91 | Shapes assumed to have formed through massive water outflow through channels originated in this area. The scenery is captured in a region called Arsinoes Chaos.

88

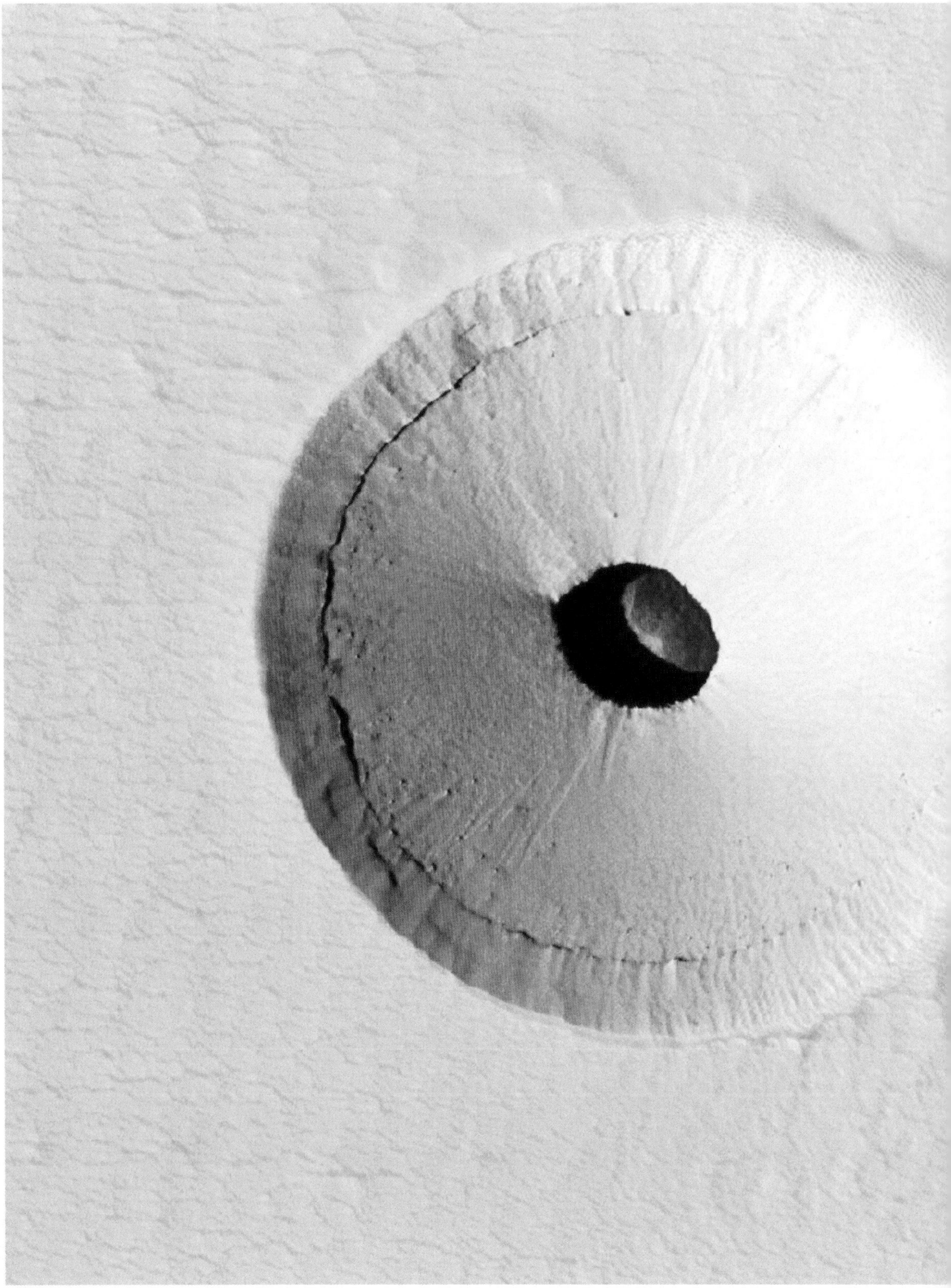

GRACEFUL FORMS

Mars's volcanoes are gigantic and evidence that the planet was geologically active, with potentially some uncovered surprises.

The planet hosts the largest volcano in our solar system, Olympus Mons. With a height of 21.9 km (13.6 mi), it displays a size almost three times the height of Mount Everest and is even visible from space.
The volcano is a massive 624 km (374 mi) in diameter located in the region of Tharsis Montes.

Page 94 | Interesting minerals are often revealed in central peaks of impact craters. They are of the highest interest to open the window further to the planet's history. This Image is captured with a diameter of less than 1km (> 1 mi).
Page 95 | Frost is lying in the shadows of these stunning dune formations. Captured in the south during Marsian autumn.
Page 97 | Fantastic ejecta of a young and spectacular impact crater.
Page 99 | Gorgeously active sand migrations and ripples over rocks in Nili Patera. The field was the first one to document a movement of a minimum of 1 m (3 ft) while monitoring over some period of time.

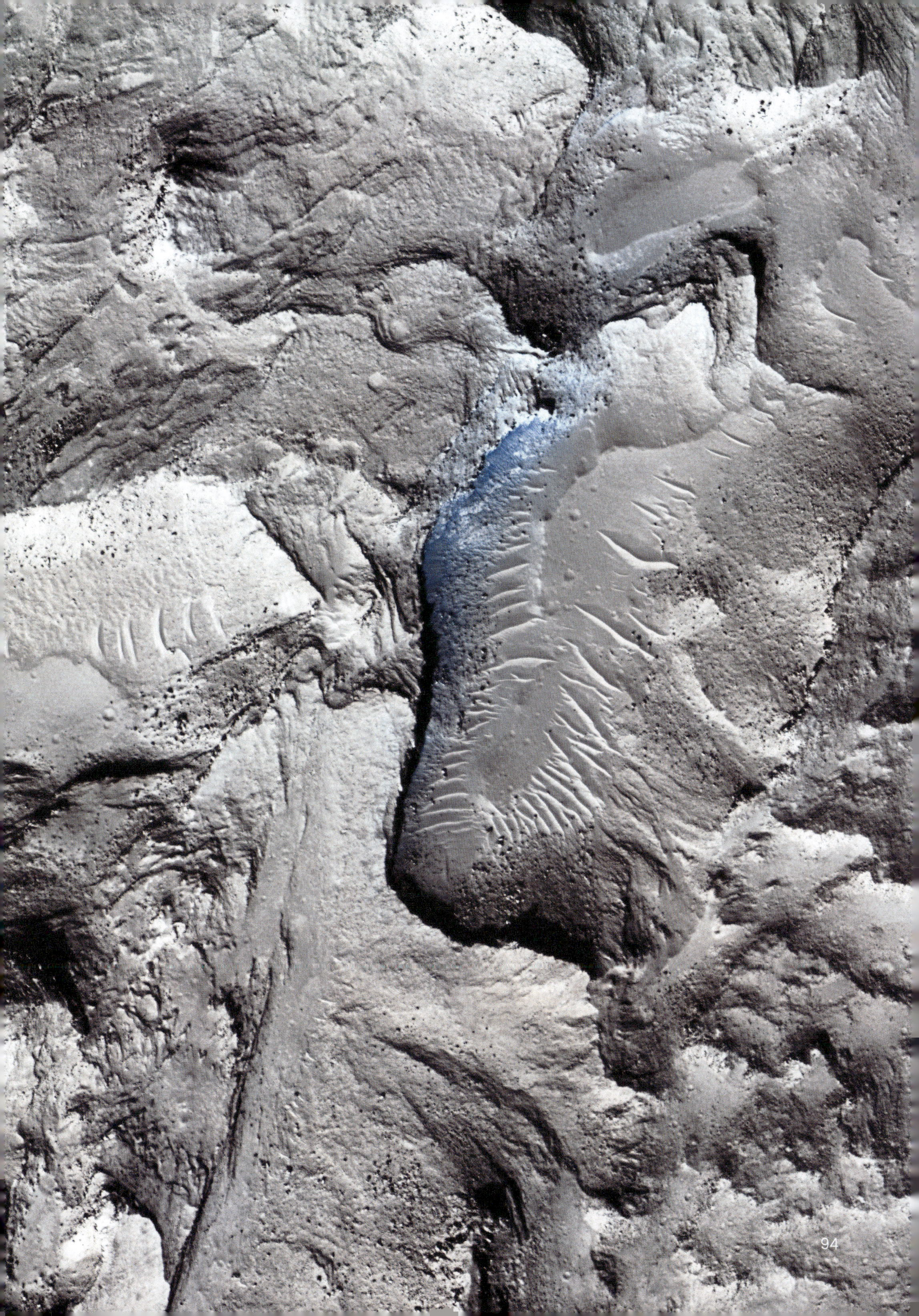

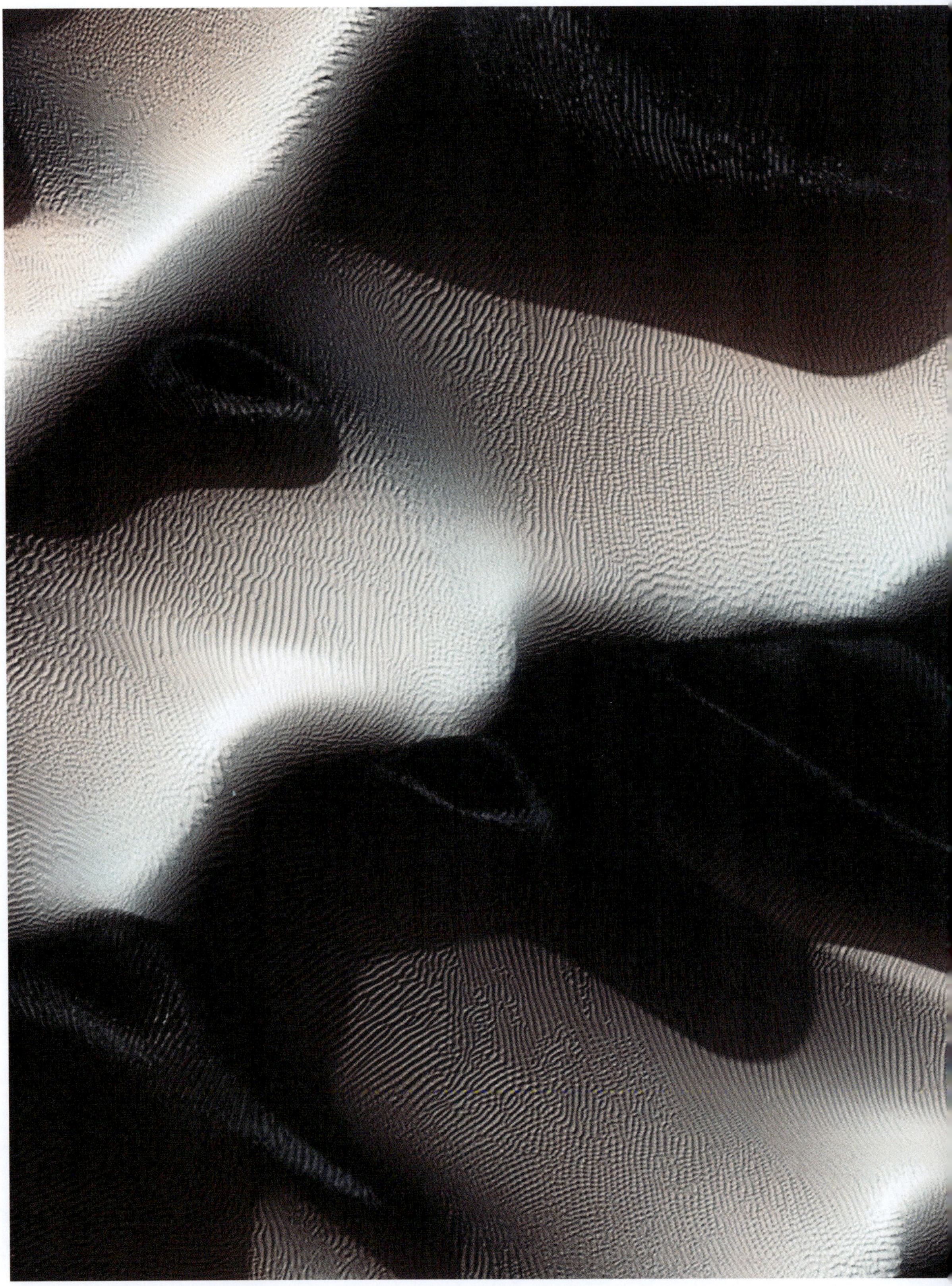

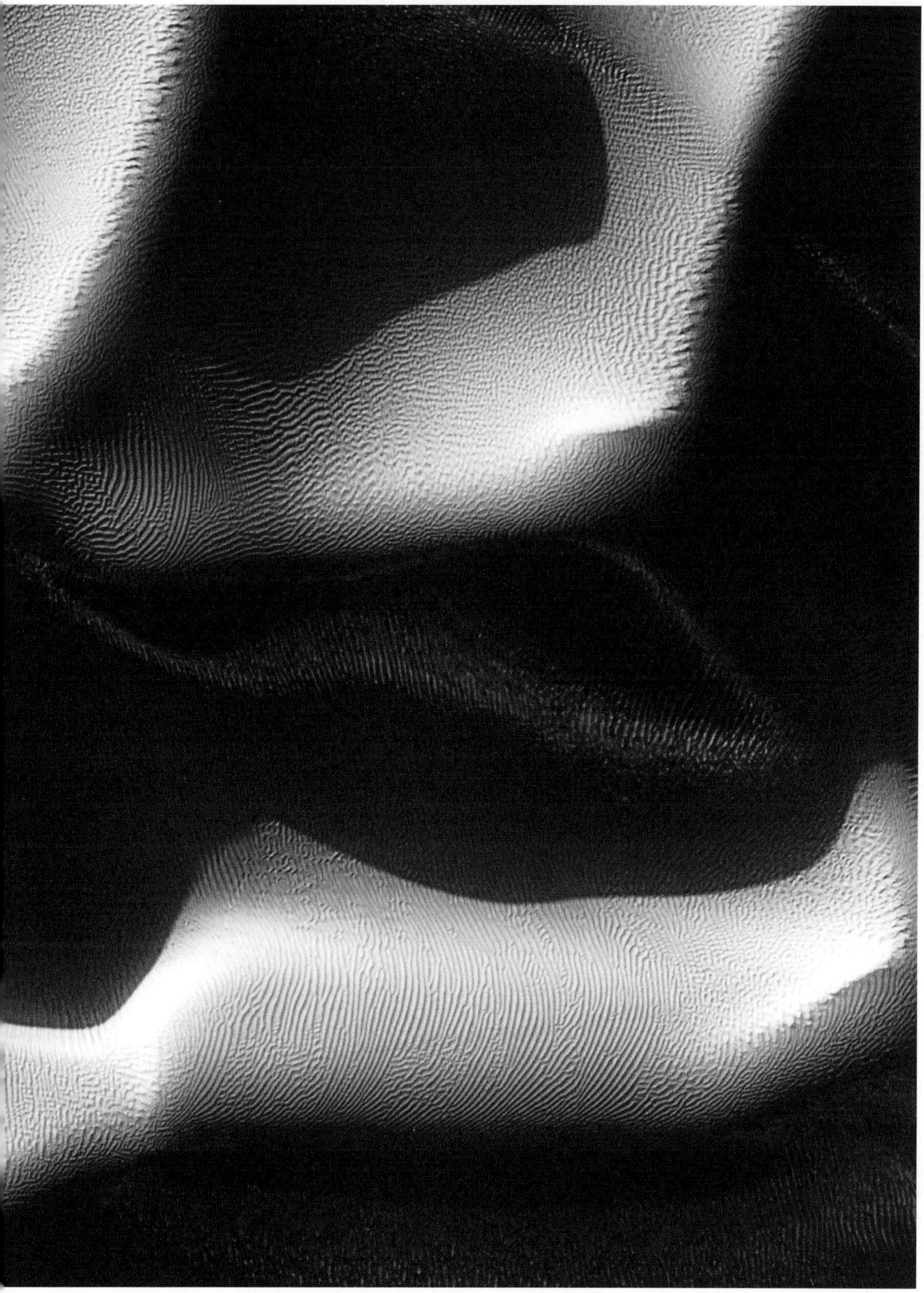

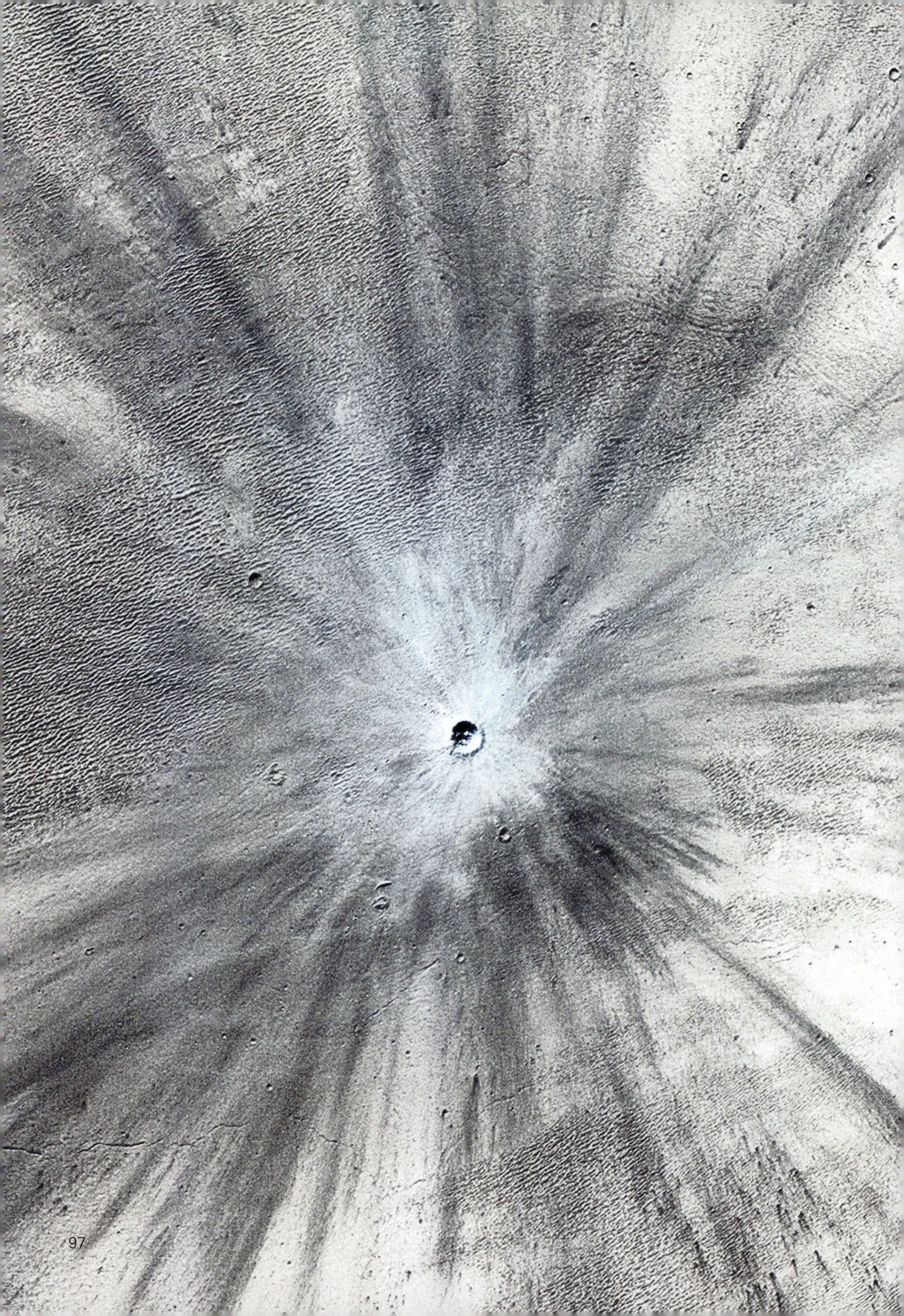

Page 101 | Lovely dunes in the north polar region. Dunes are often monitored over time to estimate if they're active or not.
Page 103 | Beautiful dark dunes located inside of the Herschel Crater. No craters have been detected on any dunes on the planet, which suggests the dunes are young and highly active.

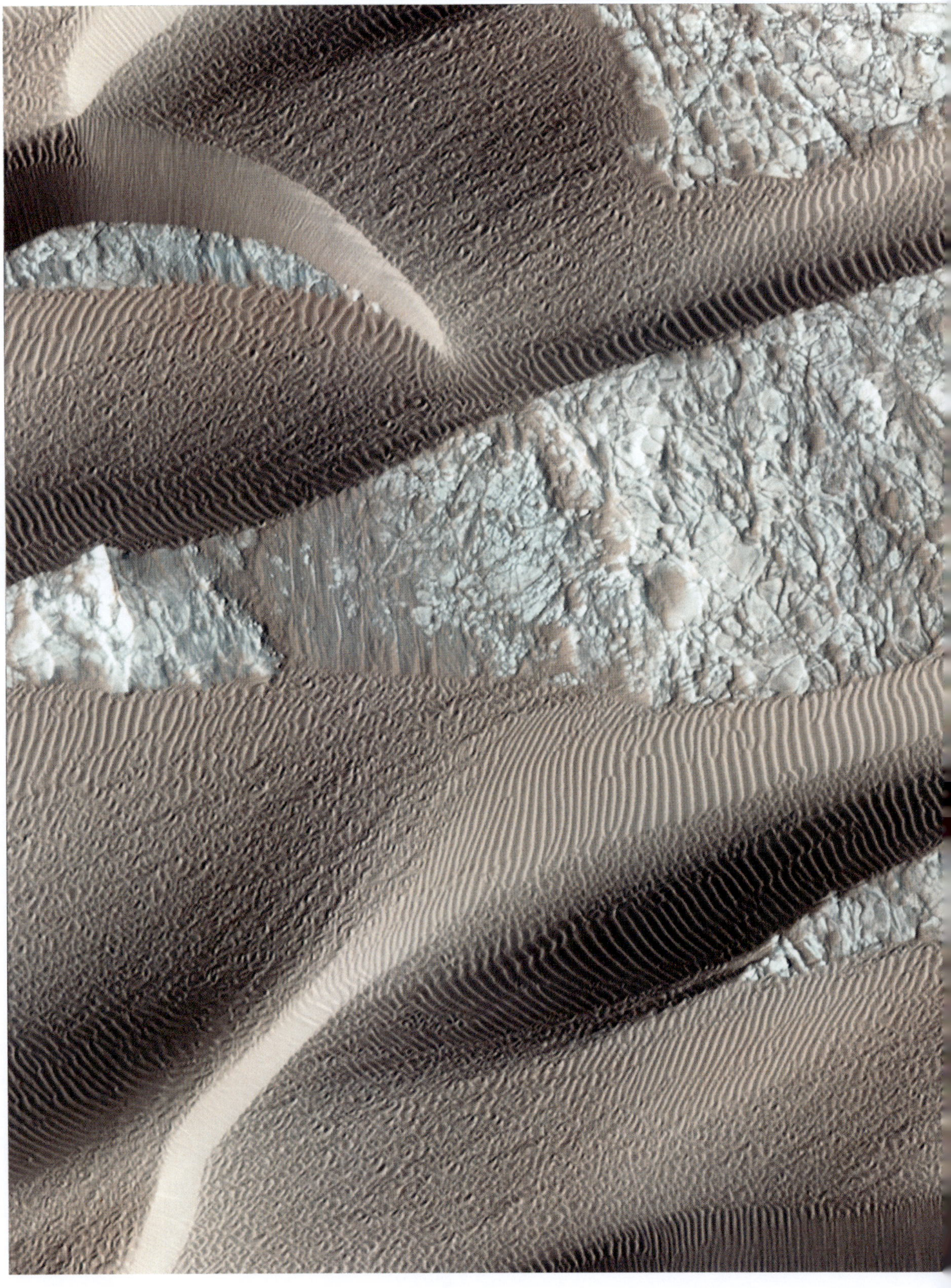

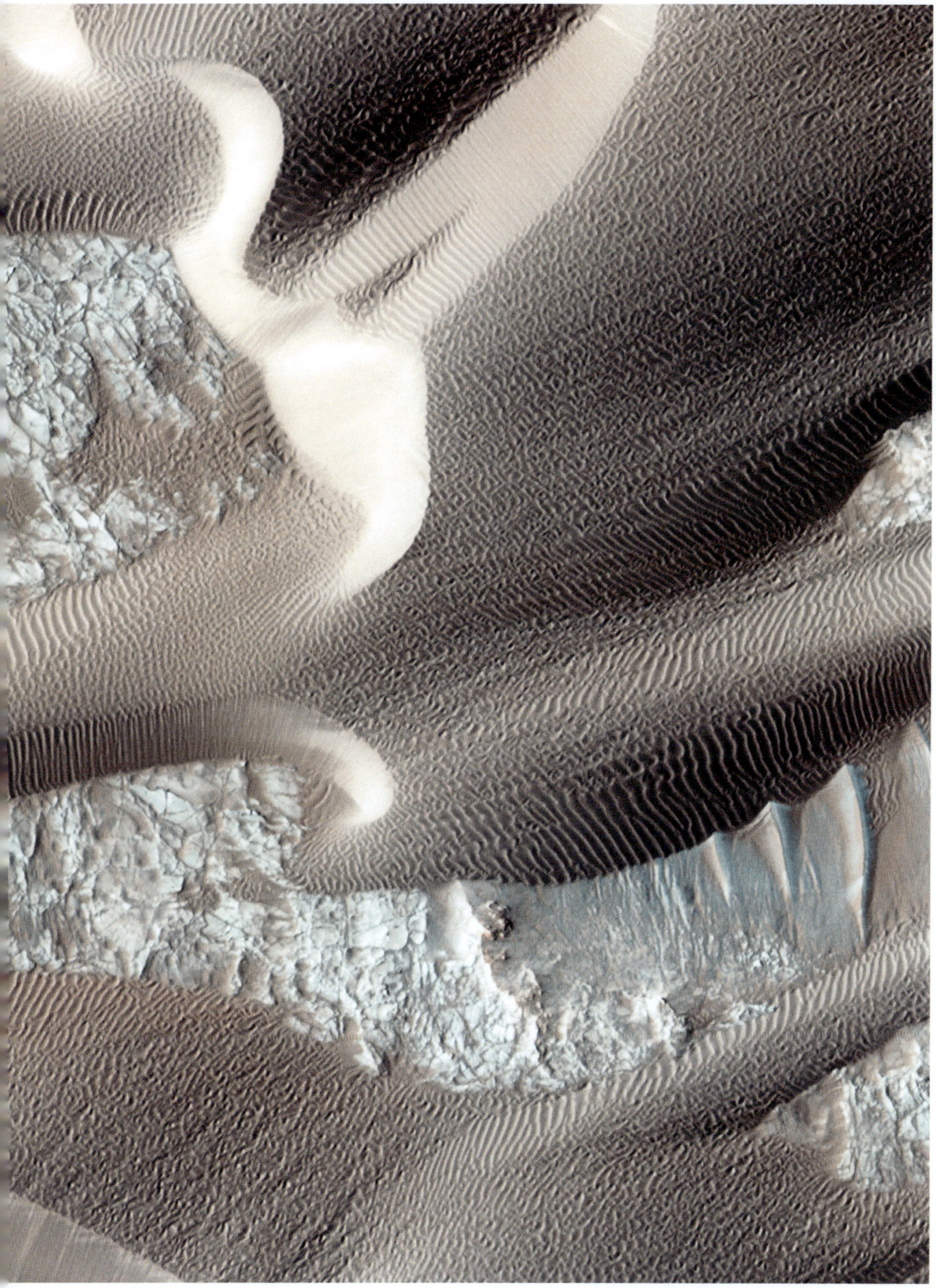

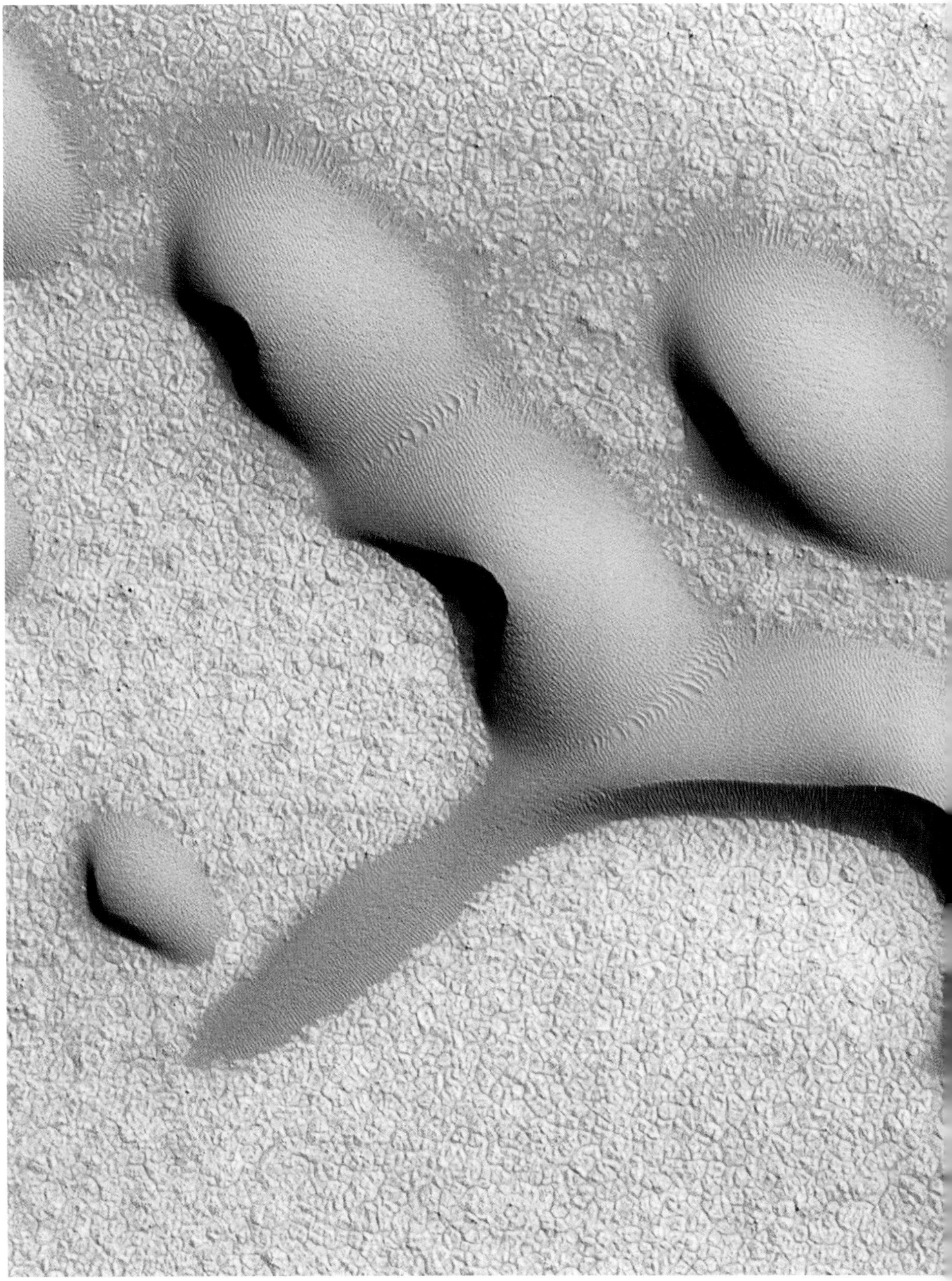

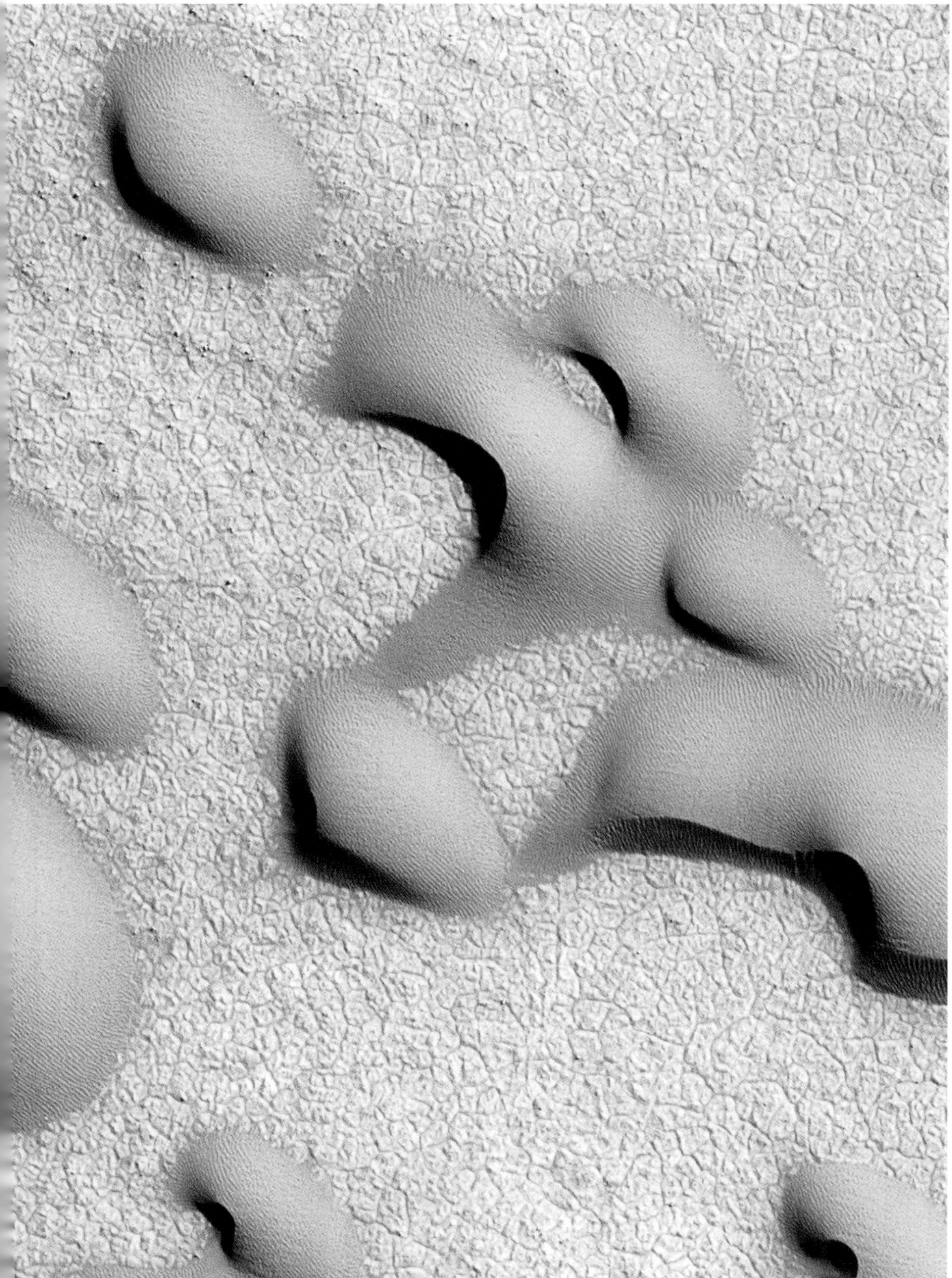

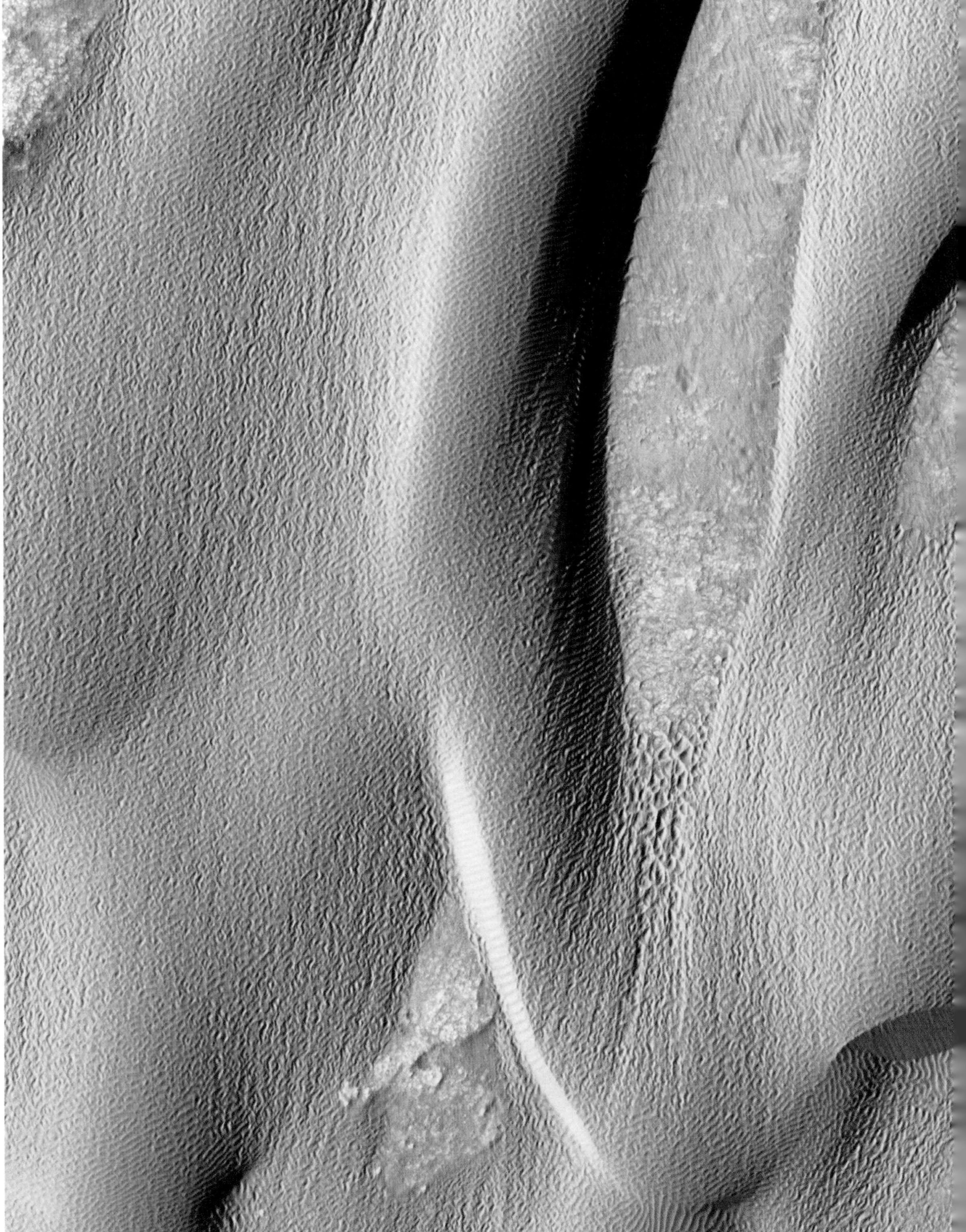

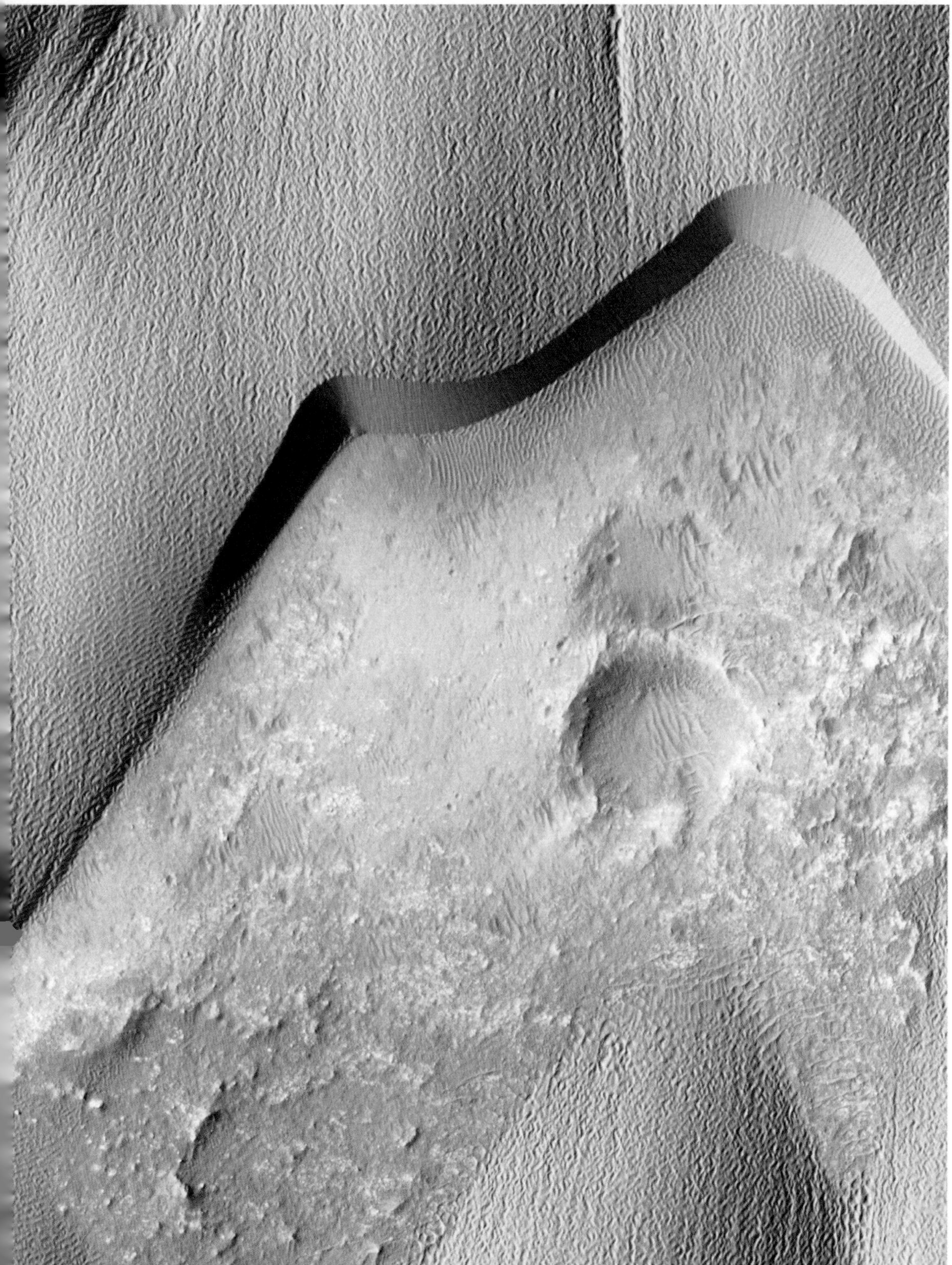

105

Page 105 | Volatile pits embedded between ejecta from the impact crater named Hale. These pits form by exploded gas due to the immense heat created by an impact.
Page 107 | Spectacular ripples and ridges among dunes.
Page 109 | This lovely depression displayed at the South Pole on Mars is likely to have formed through material that goes directly from a solid to a gas state. The planet's average temperature allows ice to form not only through water but also through concentrations of carbon dioxide.

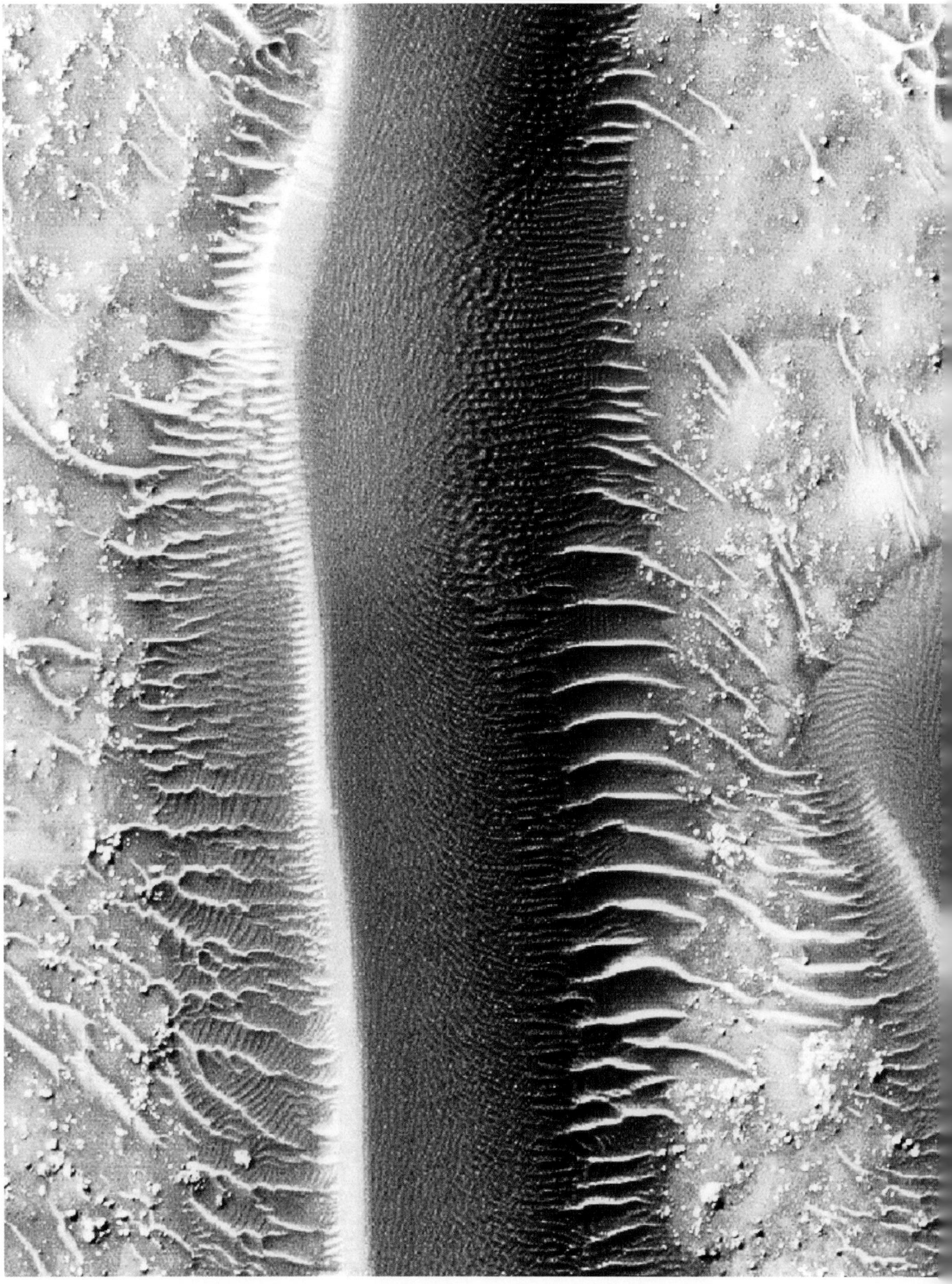

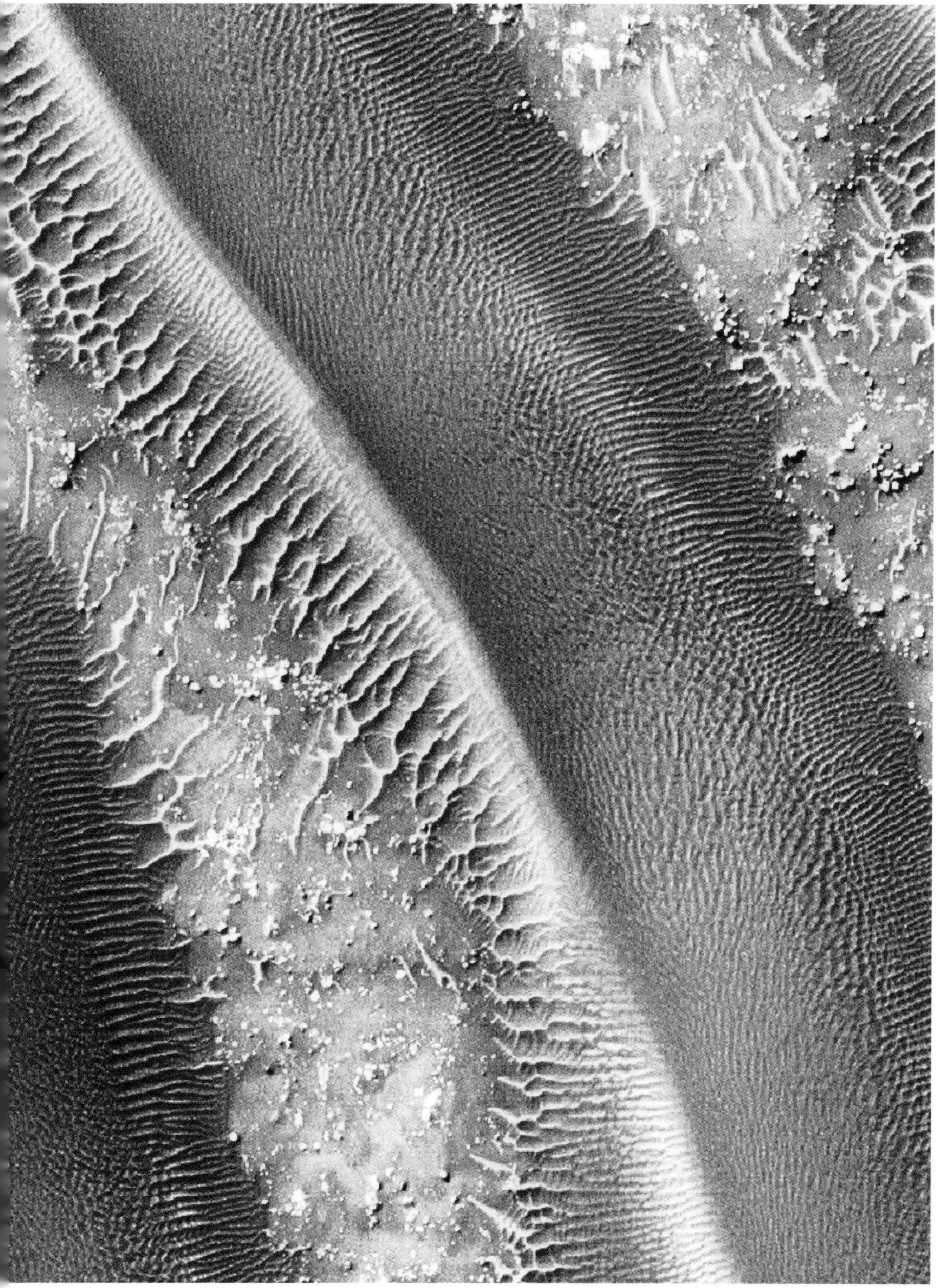

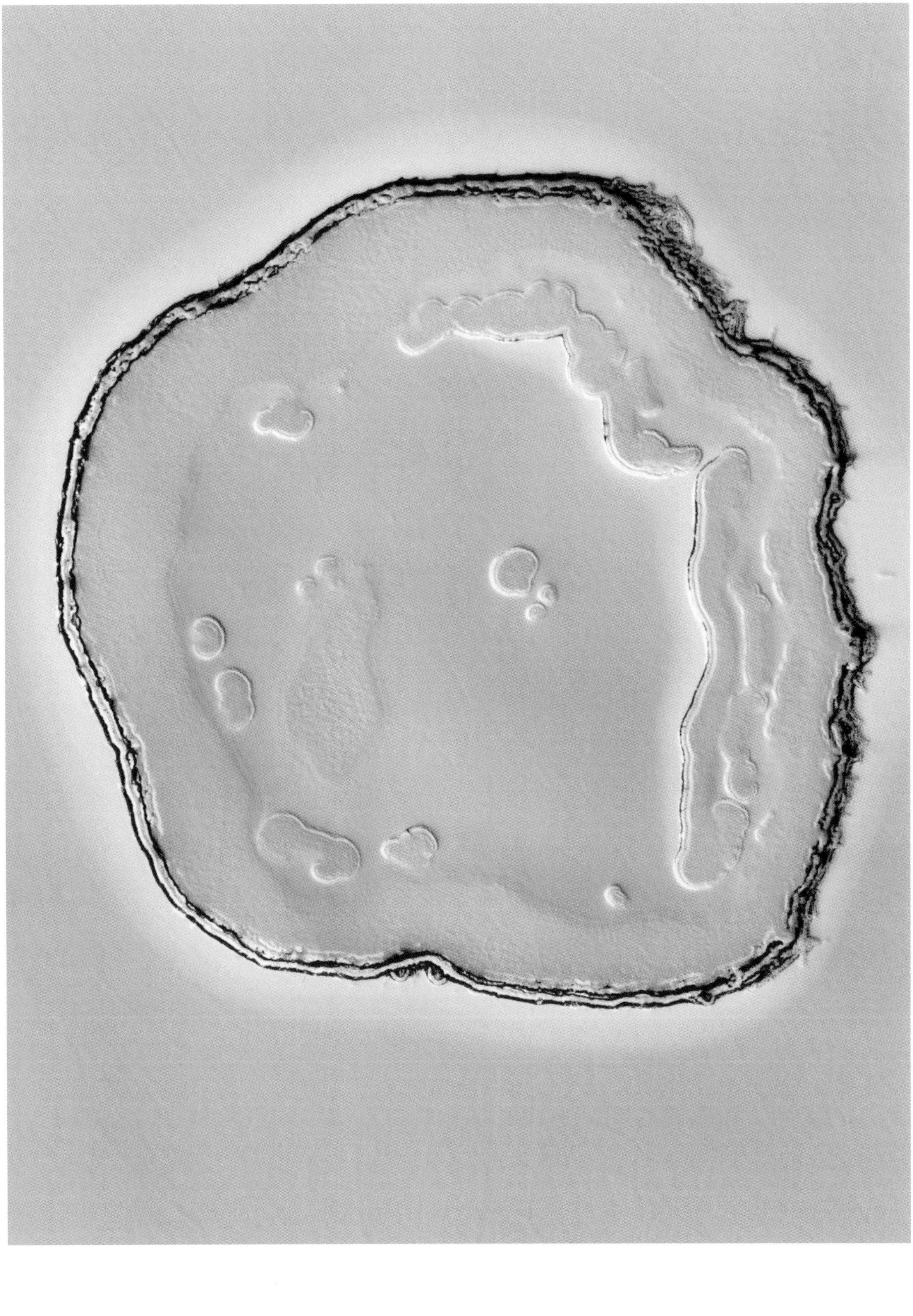

Page 111 | Image displaying gullies on an unnamed crater wall formed through its ejecta. The Crater is located in Noachis Terra.
Page 113 | Gorgeous dunes neighboring solid terrain on the floor of the Herschel Crater.
Page 115 | Beautiful crater captured in Evros Vallis. The Valley is believed to have been eroded by flowing water in the Martian past.

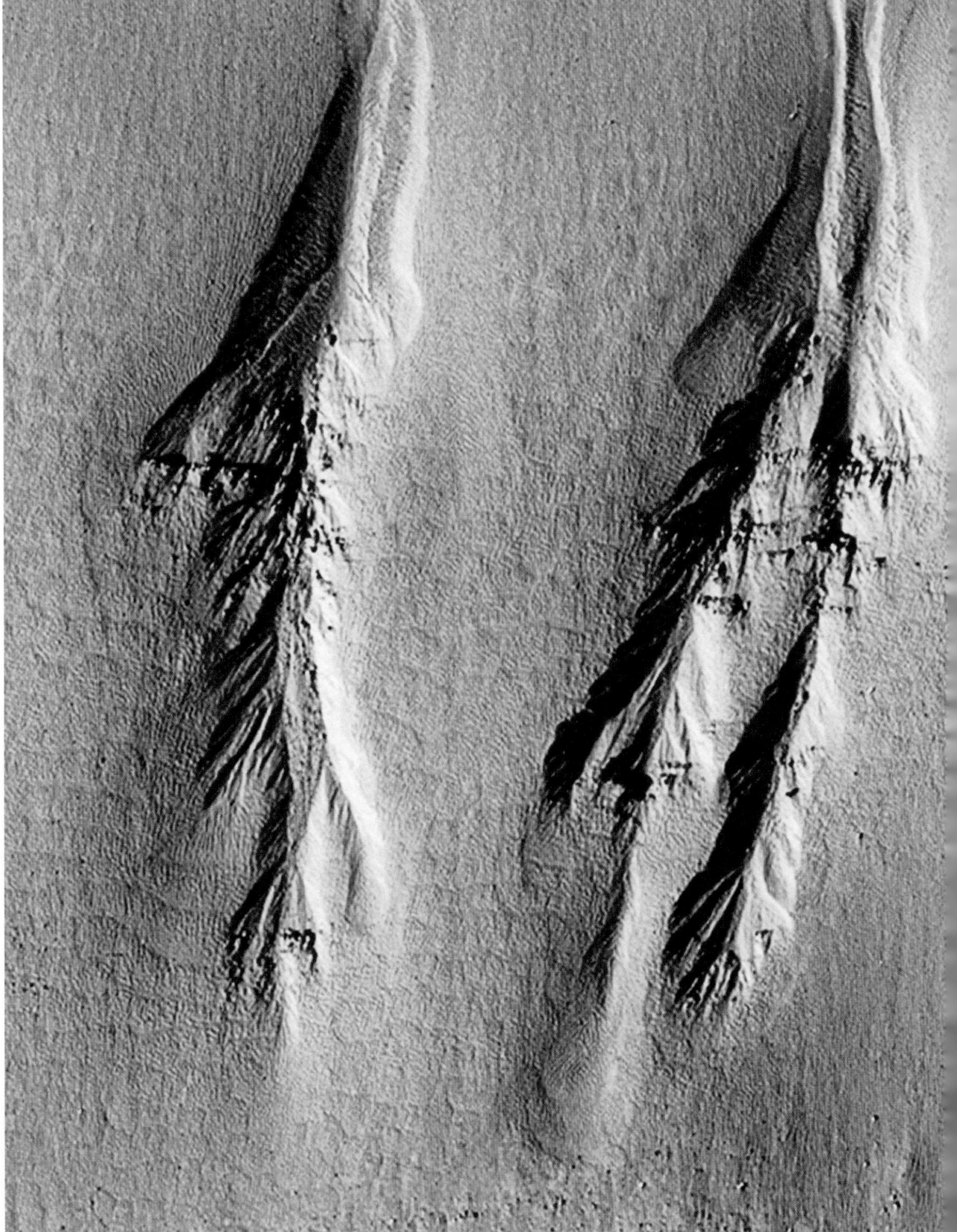

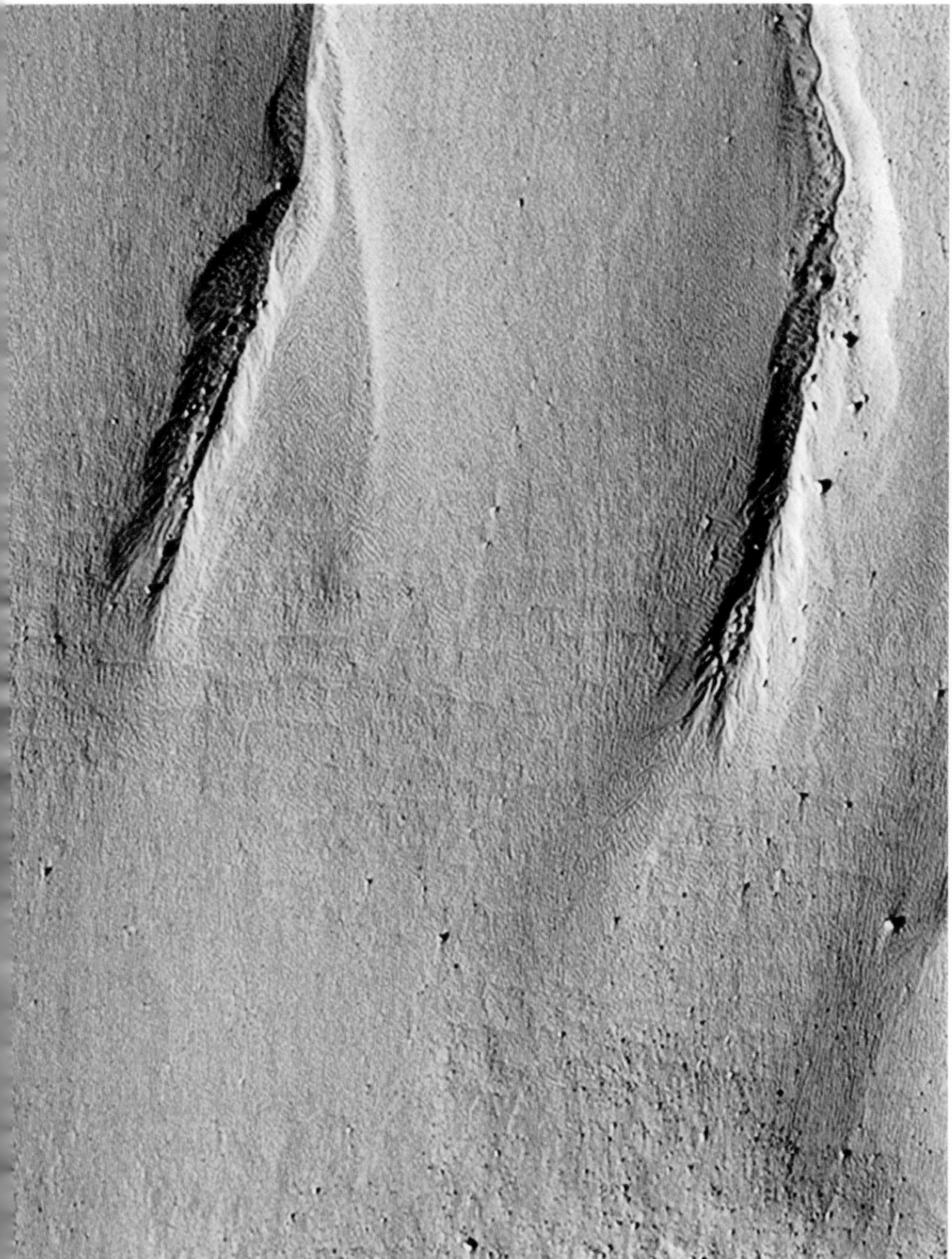

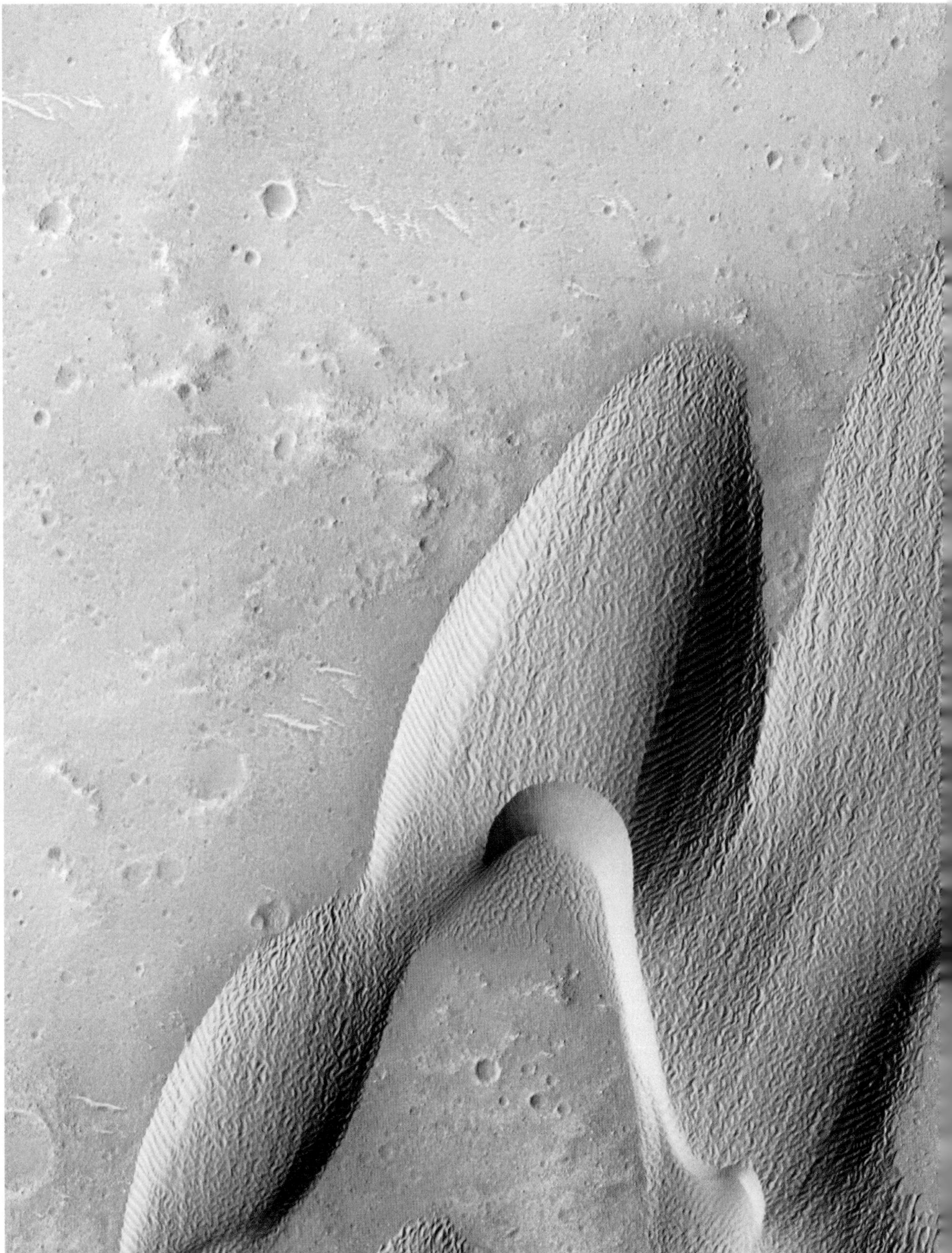

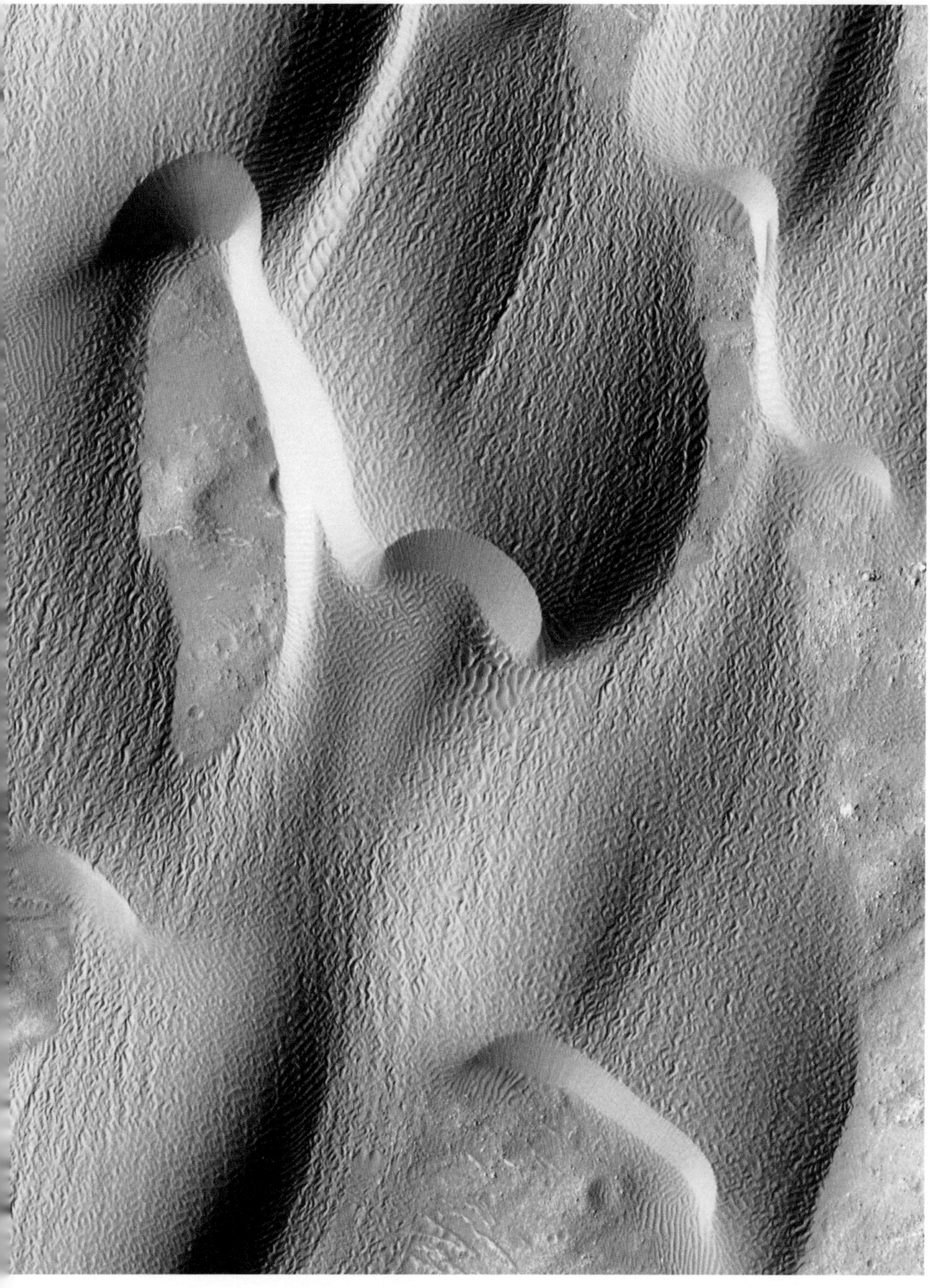

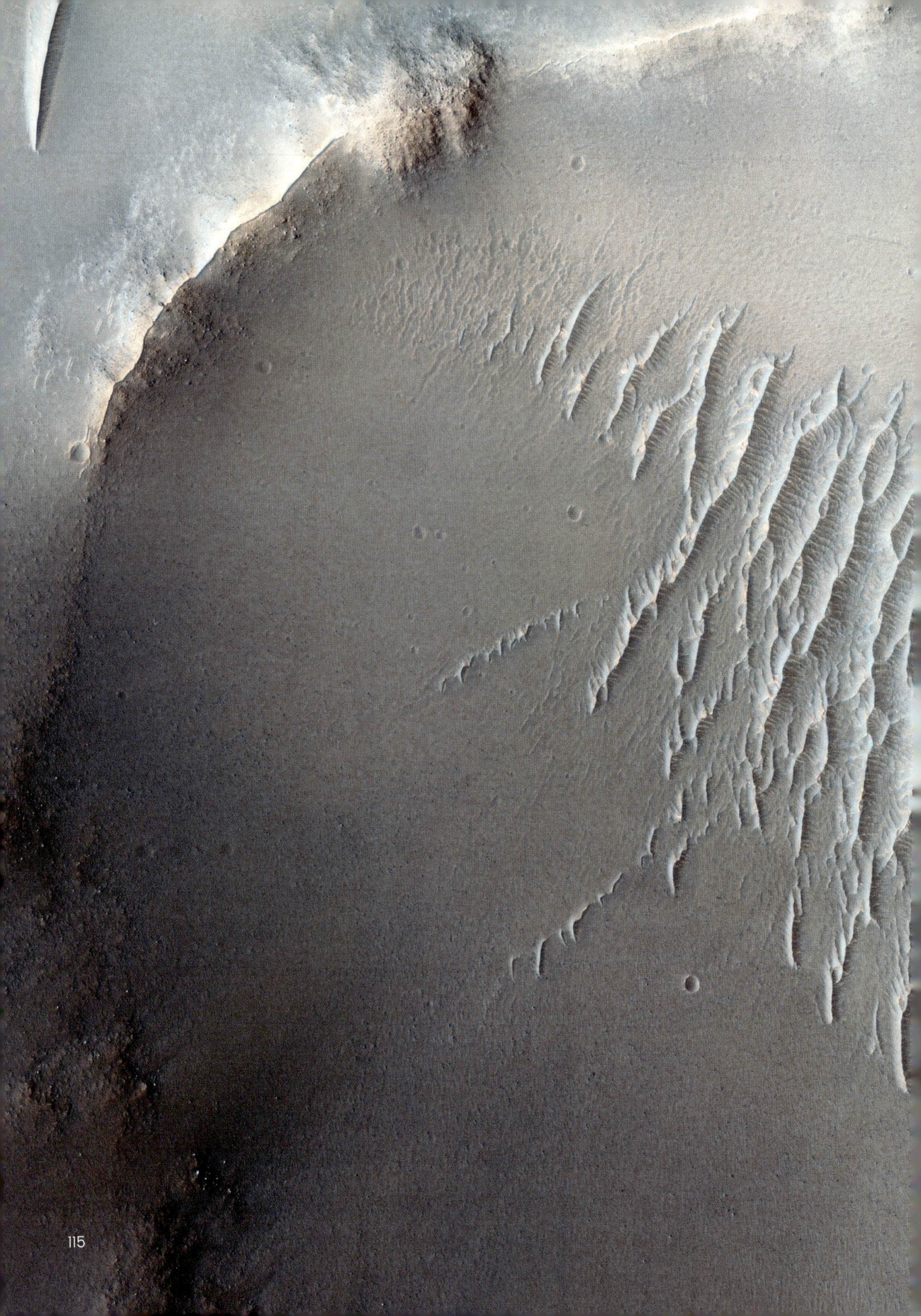

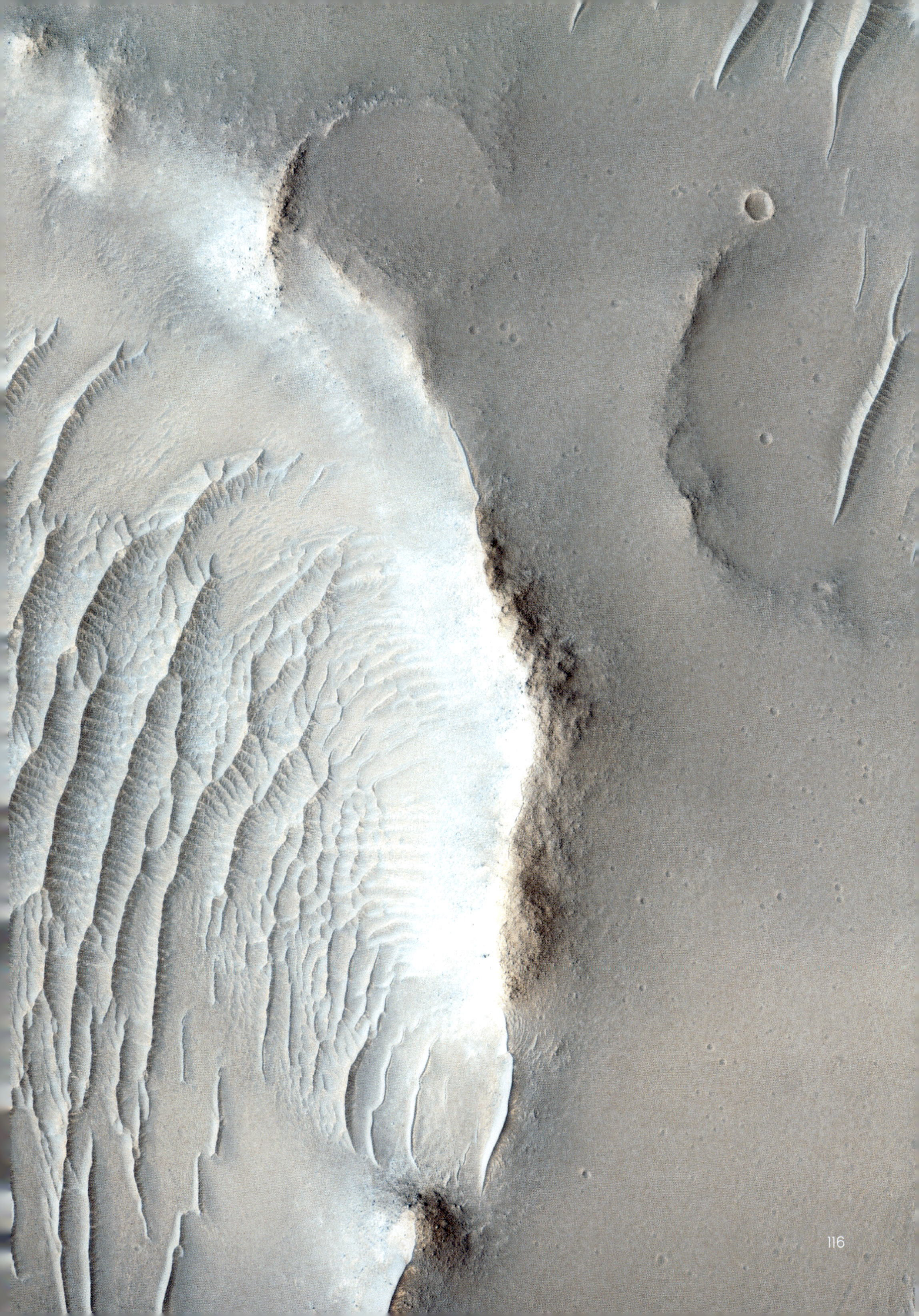

Page 118 | This magnificent sunset was captured by NASA's Rover Spirit on May 19th, 2005, after being commanded to stay awake and send this stunning sunset data.
The panorama was captured with 750nm, 530nm, and 430-nanometer color filters to generate a view similar to what a human would see but fairly exaggerated.

Mars is 1.524 astronomical units away from the Sun. One astronomical unit is the distance from the Sun to Earth. With Mars being further away, the planet has the Sun appear only about 2/3 the size as we see on Earth.

Mars offers sublime and beautiful twilight scenery, with a faint glow visible up to 2 hours before sunrise and after sunset.

———————————

All images in this book are from the archives of NASA and captured by HiRISE (High-Resolution Imaging Science Experiment) onboarding the Reconnaissance Orbiter by NASA.
The Camera can capture Mars's surface at an astonishing 30 cm/pixel and theoretically could capture a human shadow from 200-300 km (120-190 miles) above.

HiRISE is the most powerful camera ever sent to another planet from Earth.

	MARS	EARTH
Distance from Sun	149 million km (93 mil mi)	142 million km (88 mil mi)
Diameter	6790 km (4220 mi)	12'750 km (7,926 mi)
Length of Year	687 Earth Days	365.25 Days
Length of Day	24 hours 37 minutes	23 hours 56 minutes